IMAGE:
AN INTEGRATED MODEL TO
ASSESS THE GREENHOUSE EFFECT

IMAGE:
AN INTEGRATED MODEL
TO ASSESS
THE GREENHOUSE EFFECT

by

Jan Rotmans

KLUWER ACADEMIC PUBLISHERS
DORDRECHT / BOSTON / LONDON

ISBN 0-7923-0957-X

Published by Kluwer Academic Publishers,
P.O. Box 17, 3300 AA Dordrecht, The Netherlands.

Kluwer Academic Publishers incorporates
the publishing programmes of
D. Reidel, Martinus Nijhoff, Dr W. Junk and MTP Press.

Sold and distributed in the U.S.A. and Canada
by Kluwer Academic Publishers,
101 Philip Drive, Norwell, MA 02061, U.S.A.

In all other countries, sold and distributed
by Kluwer Academic Publishers Group,
P.O. Box 322, 3300 AH Dordrecht, The Netherlands.

Printed on acid-free paper

All Rights Reserved
© 1990 Kluwer Academic Publishers
No part of the material protected by this copyright notice may be reproduced or
utilized in any form or by any means, electronic or mechanical,
including photocopying, recording or by any information storage and
retrieval system, without written permission from the copyright owner.

Printed in the Netherlands

Contents

Preface	xi
Units	xii

1 General Introduction   1
    1.1 The Problem 1
    1.2 The Model 4

2 General Model description of IMAGE   11
    2.1 Introduction 11
    2.2 Description of the Various Modules 15
       2.2.1 Emission Modules 15
       2.2.2 Concentration Modules 16
       2.2.3 Climate Module 17
       2.2.4 Sea Level Rise Module 18
       2.2.5 Socio-Economic Impact Module 19
    2.3 Description of Scenarios 19
       2.3.1 Introduction 19
       2.3.2 Energy Supply 20
       2.3.3 Agriculture 24
       2.3.4 CFC Use 26
       2.3.5 Other Trends 27
    2.4 Model Deficiencies and Future Developments 27
    2.5 Comparison with Other Models 28
    2.6 Discussion 29

3 The Carbon Cycle Model   31
    3.1 Introduction 31
    3.2 Model Description 32

	3.3	Emissions Module	32
	3.4	Atmospheric Concentrations Module	34
	3.5	Ocean Module	35
	3.6	Terrestrial Biosphere Module	40
	3.7	Deforestation Module	45
		3.7.1 Introduction	45
		3.7.2 Description of the Deforestation Model	47
		3.7.3 Description of the Processes	53
	3.8	Validation and Uncertainty	62
	3.9	Results	62
	3.10	Conclusions	72
	3.11	Appendix	74

4 The Methane Module — 81

	4.1	Introduction	81
	4.2	Model Description	82
		4.2.1 Structure	82
		4.2.2 Calculation Procedure	82
		4.2.3 Notation	84
	4.3	Emissions	84
	4.4	Concentrations	90
	4.5	Results	93
	4.6	Conclusions	100

5 The N_2O Module — 103

	5.1	Introduction	103
	5.2	N_2O Emissions Module	103
	5.3	N_2O Concentration Module	106
	5.4	Results	107
	5.5	Conclusions	108

6 The CFCs Module — 111

	6.1	Introduction	111
	6.2	CFCs Emissions Module	112
	6.3	CFCs Concentrations Module	116
	6.4	Results	117
	6.5	Conclusions	121

CONTENTS

7 The Climate Module — **123**
  7.1 Introduction 123
  7.2 Model Description 125
    7.2.1 Equilibrium Response 125
    7.2.2 Transient Response 129
    7.2.3 Climate Feedbacks 133
  7.3 Results 134
  7.4 Conclusions 146

8 The Sea Level Rise Module — **149**
  8.1 Basic Trend 150
  8.2 Thermal Expansion 150
  8.3 Glaciers and Small Ice Caps 151
  8.4 Greenland Ice Cap 152
  8.5 Antarctic Ice Cap 154
  8.6 Uncertainties 155
  8.7 Sea Level Rise Potential 156
  8.8 Results 157
  8.9 Conclusion 161

9 Socio-Economic Impact — **163**
  9.1 Introduction 163
  9.2 General Model Description 164
  9.3 Quantification of Impacts for Various Sectors 171
    9.3.1 Introduction 171
    9.3.2 Coastal Defence 171
    9.3.3 Water Management and Water Supply 178
    9.3.4 Agriculture 180
    9.3.5 Energy Use 181
  9.4 Results 182
  9.5 Conclusions 191

10 Policy Analysis — **193**
  10.1 Introduction 193
  10.2 Scenario Calculations 194
  10.3 Low Climate Risk Scenario 195
  10.4 Results 196
  10.5 Delayed Response 198
  10.6 Future Worlds 200

10.7 Conclusions 203

11 Temperature Increasing Potential   **205**
  11.1 Introduction 205
  11.2 Relation between Temperature and Emissions 205
  11.3 Methodology 206
      11.3.1 Definition 206
      11.3.2 Modelling Approach 208
      11.3.3 Analytical Approach 211
  11.4 Results 215
  11.5 Conclusions 220
  11.6 Appendix 222

12 Sensitivity Analysis   **225**
  12.1 Introduction 225
  12.2 Metamodelling 226
  12.3 Experimental Design 228
  12.4 A Metamodel for the Costs of Dike Raising 230
      12.4.1 Introduction 230
      12.4.2 Input and output variables 230
      12.4.3 Specification of the First Metamodel for the Costs of Dike Raising 230
      12.4.4 Experimental Design for the First Metamodel 232
      12.4.5 Results of the First Metamodel 233
      12.4.6 Further Analysis after the First Metamodel 233
      12.4.7 Specification of the Final Metamodel for the Costs of Dike Raising 234
      12.4.8 Validation of the Final Metamodel 236
      12.4.9 Scaling Effects 239
      12.4.10 Conclusions 240
  12.5 A Metamodel for the Ocean Module 241
      12.5.1 Introduction 241
      12.5.2 First Metamodel for the Ocean Module 242
      12.5.3 Further Analysis after the First Metamodel 245
      12.5.4 Final Metamodel for the Ocean Module 245
      12.5.5 Conclusions 249
  12.6 A Terrestrial Biosphere Metamodel 249
      12.6.1 Introduction 249

        12.6.2 Various Metamodels for the Terrestrial Biosphere Module 250
        12.6.3 Conclusions 256
    12.7 General Conclusions 257

13 Discussion   **259**

References   **263**

Preface

This book is the result of a research project entitled "Reference function for Global Air Pollution/CO_2" initiated by RIVM. It deals with the description of a computer simulation model of the greenhouse effect. This model, IMAGE, tries to capture the fundamentals of the complex problem of climate change in a simplified way. The model is a multidisciplinary product and is based on knowledge from disciplines as economics, atmospheric chemistry, marine and terrestrial biogeochemistry, ecology, climatology, and glaciology. This book might be of interest for any one working in the broad field of climate change. Furthermore, it can be useful for model builders, simulation experts, mathematicians etc. A PC version of the model will become available free of charge. Requests can be sent to the author.

Although being the only author of this book, I could never have written it without the help of many other people. First of all I would like to thank Koos Vrieze, originally a colleague at RIVM, later my professor. Without his inspiring attitude I would have never finished this thesis. I am also very grateful to RIVM for giving me the opportunity to write this thesis. I owe many thanks to Hans de Boois and Rob Swart for their support and assistance during the research. Furthermore, I would like to thank my trainees who have substantially contributed to the contents of this book. Especially I would like to thank Greet van Ham, who helped me to perform a thorough sensitivity analysis with the computer model. I also want to express my attitude for the help of Michel den Elzen, who assisted me with great devotion in developing and improving parts of the model. Special thanks to Martin Middelburg and André Berends for drawing the great number of pictures. I also wish to thank Marlies Haenen for the many hours she spent on transforming the rough version of the document into a perfect-looking one.

Last, but not least, I would like to thank Inge and my parents for their support and patience throughout.

Units

$°C$	degrees Celsius
K	degrees Kelvin
ppm	parts per million by volume
ppb	parts per billion by volume
ppt	parts per trillion by volume
Gt	gigatons (10^{15} gram)
GtC	gigatons of carbon
Tg	teragram (10^{12} gram)
Mkg	million kilogram (10^{9} gram)
λ	lambda
Wm^{-2}	watts per meter square
Mha	millions of hectares
kg	kilogram
TgN	teragram of nitrogen
Dfl	Dutch florins or guilders
km	kilometer

Chapter 1

General Introduction

1.1 The Problem

The earth's energy balance is dominated by atmospheric heat trapping; about 70% of of the incoming solar energy is absorbed. The absorbed energy is re-emitted at infrared wavelengths by the atmosphere and the earth's surface. Most of the surface radiation is trapped by clouds and greenhouse gases, and returned to the earth. This phenomenon is called the greenhouse effect, inducing an average temperature of the earth at the surface of +15 $°C$ rather than the -18 $°C$ it would have without the heat trapping. This greenhouse effect is essential for the existence of life on earth, having made it possible for millions of years.

Through this process of radiative absorption greenhouse gases — of which carbon dioxide (CO_2), methane (CH_4), nitrous oxide (N_2O), and chlorofluorocarbons (CFCs), together with water vapour (H_2O) and ozone (O_3) are the most important — play a major role in determining the earth's climate.

Since the industrial revolution mankind has caused a considerable increase in the atmospheric content of these greenhouse gases. The primary cause of the increased concentrations is the large-scale and worldwide use of fossil fuels to generate heat, power, and electricity for a growing world population, as well as the changes in land use, especially for agriculture. For more than a century the use of fossil fuels has been on the increase and this development is still continuing. Since the beginning of this century the atmospheric concentration of CO_2 has increased by about 25% . The concentrations of some other greenhouse gases have increased by even larger factors.

Dependent on the assumptions and definitions used it can generally be said that the energy and industrial sectors cause about 75% of the enhanced greenhouse problem, while about 25% is caused by the agricultural sector, including deforestation. In this way mankind has become a dominant factor in the greenhouse effect.

There is virtually no doubt among scientists that increasing concentrations of greenhouse gases will raise the earth's temperature. There is only controversy about the magnitude and the time frame of the temperature increase. Measurements from bubbles of air trapped in ice cores of the Antarctic ice sheets show that, over the past 160,000 years, there is a very close correlation between past climatic changes and fluctuations in the greenhouse gas concentrations of carbon dioxide and methane. It is not clear, however, whether the greenhouse gas variations caused the climatic changes or vice versa.

In the past 100 years the global average surface temperature has increased by about half a degree celsius. Although it seems obvious to account for the temperature increase on the basis of the enhanced greenhouse effect, the evidence is not yet conclusive. One of the things that cannot be explained is the irregular course of the temperature record: a rapid warming until the end of the Second World War, then a slight cooling through the seventies, and a period of rapid warming since then. That does not correspond with the steady warming that might be expected from a steady buildup of the greenhouse gases.

The future temperature curve will depend on the emissions of trace gases, the resulting concentrations, and the climatic effects these concentrations will bring about. Growth in fossil fuel use, the adoption of alternative energy sources, the rate of deforestation, and policy measures will strongly affect future emission pathways. These future emission levels will, together with a whole complex of feedback mechanisms, determine the resulting concentrations, among other things. Increasing concentrations and temperature rise may trigger both positive and negative feedback mechanisms, such as the increased uptake of CO_2 by plants and oceans, or increased release of CO_2 by organic matter in soils, or increasing CH_4 release through methane hydrates and methane locked up in Arctic permafrost. In spite of these uncertainties, a doubling of the CO_2 concentration is expected around the middle of the next century. The CO_2 equivalent concentration, which expresses the combined total radiative forcing effect of all trace gases compared to CO_2 alone, is expected to double even sooner.

The climatic implications of a doubling of atmospheric CO_2 are computed by mathematical climate models. These models are used because history

1.1. THE PROBLEM

gives us no clear quantitative answer, and the complicated climate system cannot be reproduced in a laboratory. These General Circulation Models (GCM's) consist of mathematical expressions for the ocean-atmosphere system. In such a model the atmosphere is represented as a three-dimensional grid with an average horizontal spacing of several hundred kilometers, and an average vertical spacing of several kilometers. The models calculate equilibrium temperature, pressure, wind speed, humidity, soil moisture and other variables, based on an abrupt doubling or quadrupling of atmospheric CO_2. Running such GCM's costs immense amounts of computer time even on the fastest supercomputers (Schneider, 1989).

These climate models suggest that the earth is now warmer than the observational record indicates, and are inconsistent in their simulation of regional-scale average temperatures. Simulated precipitation is also geographically inconsistent among the various models. The inconsistencies of the models are due to their coarse resolution and their crude treatment of the oceans. However, smaller-scale portrayals, while desirable, increase model computational requirements exponentially.

Despite their limitations, models are the only means for estimating future climate change. The results of the most recent GCM's are in agreement that a doubling of CO_2 in the atmosphere will lead to an average surface temperature increase of 1.5 $°C$ to 4.5 $°C$. This estimate for the coming century is comparable with the increase in temperature of roughly 5 $°C$ since the peak of the last ice age 18,000 years ago, with this difference: it will occur between 10 and 100 times faster (Schneider, 1989).

These predicted temperature increases will be larger at polar latitudes than at temperate latitudes and smallest at tropical latitudes. Additionally, precipitation will increase globally by 7% to 15%.

In spite of the discrepancy between the model predictions and the measured global temperature trend so far, there is a growing scientific consensus that the enhanced greenhouse warming has already appeared, and that it will be detected within one or more decades.

Obviously the greenhouse phenomenon will have a profound effect on ecological and social structures. One of the major threats is a sea level rise, as a result of the thermal expansion of the oceans, the melting of glaciers, and the net effect of the possible melting of the Greenland ice cap and the accumulation effect in Antarctica. Sea level rise could endanger many low-lying coastal settlements and ecosystems, and might cause erosion, salt intrusion and seepage.

Water resource management, currently a problem in many nations, may become even more problematic. Current policies may have to be modified in the areas of water storage, flood and erosion control.

Practically every other important natural resource and need for human welfare, such as agriculture, shipping, forestry, nutrition, health, and natural ecosystems could be affected by the greenhouse effect, with outcomes that are currently uncertain. Some areas may benefit from the greenhouse effect.

Clearly these direct effects of climatic change could have powerful economic, social and political consequences. But the economy and society will also be indirectly affected owing to the effects on the food supply, world timber production and the energy supply. Both for the developed countries, and for the developing countries, especially for the rapidly increasing population of the Thirld World, where ecosystems are often vulnerable the risks are very high. However, even in a highly developed country with a complex social and economic structure, the environmental and economic effects may also be profound. The greenhouse issue may even lead to political instability. In the face of this array of threats, three kinds of responses could be considered: first, technical measures to counteract climatic change; secondly, adaptation, often with little or no attempt to anticipate damages or prevent climatic change. The third category of response is prevention: curtailing the greenhouse gas buildup. Energy conservation measures, alternative energy sources or a switch from coal to natural gas; whatever action we choose to take, the greenhouse effect is a permanent part of living on our fragile planet.

In summary it can be concluded that humanity is performing a large-scale geophysical experiment, not in a laboratory or on a computer, but on the planet earth itself. The outcome of the experiment should be evident within decades.

1.2 The Model

In 1984, the RIVM initiated a project entitled "Reference Function for Global Air Pollution /CO_2". Within this framework global modelling was defined as a quintessential activity. As a kind of feasibility study, it was planned to build a prototyping model for the greenhouse problem. During the period from June 1985 to February 1986 I was seconded to RIVM as a student, and in this period I developed this prototype. The main purpose of the prototype was to develop a tool that could give a broad overview of the complex greenhouse problem, by coarsely combining and aggregating diffuse information from various disciplines. The prototype comprised many input-output

1.2. THE MODEL

relationships, which underly complex model formulations. This model was mainly based on an extensive study of the literature (Rotmans, 1986), and was implemented on a PC, making use of the simulation language DYNAMO (Pugh, 1983). As a result of the satisfactory results obtained with the prototype, it seemed worth the effort to continue with a globally model of the greenhouse issue. The prototype marked the end of the first phase.

The second phase started in the course of 1986, with the initiation of IMAGE: the Integrated Model to Assess the Greenhouse Effect. The primary objective of IMAGE is to create a comprehensive picture of global climate change, by integrating the separate components into a synthetic framework. Although, in the literature, more abstract definitions of integrated environmental models are given (Hafkamp, 1984, and Brouwer, 1987), here a pragmatic definition of an integrated environmental model is used. According to Olsthoorn (1987) and Aldenberg (1988), an integrated environmental model is defined here as a model that couples air, soil, and water compartments, and integrates the interactions between divers environmental compartments. Such an integrated model considers the whole cause-effect relationship, from the arising of the pollution or the taking of a measure, to the ultimate ecological or socio-economic effect. Such an integrated model adds nothing to the separate compartment models, but is merely a modelling integration of knowledge and expertise from the separate scientific disciplines. (Boois, de, et al., 1989)

IMAGE comprises a concatenation of autonomously functioning models, called modules, each module representing and covering a particular scientific field. Following Hettelingh (1989) a module is interpreted as a set of intra-related variables oriented towards a single discipline, that represent the phenomena of a subsystem and that are interrelated to the variables of another module.

The atmosphere, terrestrial biota, and the ocean are the compartments included in IMAGE. Furthermore, the main interactions between these compartments are integrated. Within these compartments the following processes can be distinguished: emissions, atmospheric chemical processes, radiative perturbation, and sea level rise. To model these processes, as well as their causes and effects, IMAGE makes use of separate aspect-compartment models in the sphere of such scientific disciplines as the world economy, atmospheric chemistry, marine and terrestrial biogeochemistry, ecology, climatology, hydrology, and glaciology.

These elaborate aspect-compartment models need to be reformulated, and remodelled. Inevitably they have to be simplified, by applying the simplification tools of parametrization, omission of minor contributions, and approximation of nonlinear systems. Thissen (1978) has given an elaborate review of useful techniques for simplifying complicated models.

Besides this reformulation and remodelling aspect, the methodology of coupling also involves the definition of one single mathematical concept. Specifically, this means that, from a mathematical viewpoint, the broad collection of models, which can be subdivided into discrete/continuous, deterministic/stochastic, dynamic/steady state, etc., has to be reduced to one, unambiguous mathematical formulation and system (Aldenberg, 1988). Furthermore, a mutual harmonization of such aspects as content is also necessary. This implies one conceptual framework, with uniformity with respect to time and space scales, aggregation levels, use of quantities, time step, data, etc.

With respect to these aspects IMAGE meets the requirements of the definition of an integrated environmental model as given above. Following Aldenberg (1988), these integrated environmental models can in turn be subdivided into: sequentially coupled (asynchronously running models, on a variety of computers using different languages), simultaneously coupled (simultaneously running models, on a variety of computers using different languages) and fully integrated (simultaneously running models, using one computer and one language). IMAGE falls within the last category, and is a fully integrated environmental model. This has the advantage that the full integration aspect allows feedback mechanisms to be implemented rather easily, compared to the other two types of integrated model. IMAGE runs on a SUN 4 SPARC Workstation using the simulation language ACSL (Advanced Continuous Simulation Language), which has the advantage that ACSL computer code resembles the original model formulation. Separate modules are running on an IBM PS2/70 computer, again using ACSL.

In the second half of 1990 the second phase will be completed with the incorporation of a network of feedback mechanisms. Although some feedback loops have already been incorporated into IMAGE, this has been done in a provisional way. It is intended to include as many potential interactions and feedbacks as possible, and to make these into a coherent whole. As a final step of the second phase, IMAGE will be documented and a rough version of IMAGE will become available, free of charge, for any interested person or institute.

The third phase encompasses a further regionalization at every level, and the development of a menu-driven, fully interactive version of IMAGE, with a projected time schedule of five years.

1.2. THE MODEL

IMAGE belongs to the class of so-called global models (Soest, van, et al., 1988). Up to 1972, when the Report of the Club of Rome appeared, global modelling was not considered to be a serious activity. Today, more than twenty years after the foundation of the Club of Rome, it is now widely accepted that such global, empirical scenario models are a powerful tool for analysing long-term decision problems. In spite of this there is still a dense aura of scepticism around global modelling. One of the principal reasons for this scepticism is the complexity and unmanageability of the models, in combination with the accumulation of uncertainties and sensitivities. Pestel (1988) for instance, prefers the use of mental models, instead of building huge and monolithic models, which are extremely difficult to understand.

Generally, the predictive force of these models is rather limited. Rather than prediction tools, global models are neither more nor less than instruments which can amplify our insights into the present and future driving forces behind our complex social structures. In Soest van, et al. (1988) Meadows illustrates this by comparing these models with maps: "They don't tell you where to go, but if you know where to go, you can select the best route, which depends upon your goal. The chart (or equivalently the model) gives you but possibilities (options)."

This sceptical attitude holds particularly for IMAGE, especially since IMAGE is a quantitative concatenation of interacting numerical variables, according to mathematical formulations, about the hypothesis of the greenhouse effect. Consequently, if the greenhouse hypothesis were to be rejected in the future, the fundamentals underlying IMAGE would be tarnished. Therefore, although the integrated approach is conceptually attractive, it is a disputable one, because of the sequence of failures and uncertainties consequent on the modelling of the individual disciplines.

It can be questioned what the use is of such a global integrated environmental model as IMAGE. IMAGE tries to capture as much as possible of the cause-effect relationship with respect to climate change, based on an interdisciplinary approach. This leads to a better understanding of the interrelations between the different scientific skills with respect to global climate change. In this way policy agencies can be offered a concise overview of the quantitative aspects of and insights into the greenhouse problem. Additionally, such an integrating instrument can strongly increase awareness among different societal groups. Thus the educational and instructive value of IMAGE is a key element.

Besides these aspects, uncertainties or crucial gaps in current knowledge in the field of global climate change can be identified, because of the added value of the integrated approach, which can yield insights that scattered information cannot offer. Furthermore, such an integrated model reveals weaknesses in discipline-oriented models. Finally, the integrating tool enables the evaluation of long term climate strategies. The prognostic character of IMAGE makes it possible to calculate policy options, by defining scenarios. In the long run it is meant to be an interactive tool for different kinds of users, such as policy makers, investigators, students, etc.

As a matter of course, uncertainty and sensitivity analyses are of vital importance in determining the essential features and revealing the weaknesses of IMAGE. In Jansen et al. (1990) an enumeration of methods and techniques for the performance of such analyses is given. Several analyses on IMAGE have been carried out; sensitivity analyses, making use of the techniques of metamodelling and experimental designs (Rotmans and Vrieze, 1990, Ham van, et al., 1990). According to this method the relationship between the inputs and outputs of the simulation model is modelled through a regression model, also called a metamodel. The use of experimental designs allows experiments to be carried out in an efficient and effective way.

In addition, a preliminary uncertainty analysis has been performed, using the software package PRISM. PRISM incorporates the features of Latin Hypercube Sampling and metamodelling (Gardner et al., 1983, Lammerts, 1989). Finally, the method of group screening has been utilized, by which very many variables can be investigated and, hopefully, a few really important factors can be detected (Bettonvil, 1990, Bettonvil and Rotmans, 1990).

However, only modules of IMAGE have been subjected to these analyses. Unfortunately no overall analysis has been achieved on IMAGE as a whole, primarily because these sensitivity and uncertainty methods are very time-consuming. Another reason is the continually updating status of IMAGE, which makes it rather difficult to spend much time on analysing the model. For instance, the mere evaluation of the carbon cycle took more than six months.

Another delicate, unsound point which applies to global integrated environmental models, but to IMAGE in particular, is the limited possibility of verification and validation. One of the things that can at least be done is the comparison of historical trends of state variables in the model with measurements. Thus, global atmospheric concentration trends of greenhouse gases, simulated with IMAGE, are compared to measured global concentrations. For CO_2, which has been measured since 1958 at Mauna Loa, this verification time-span is more than 30 years. Furthermore, the observed global

1.2. THE MODEL

surface temperature increase of about 0.5 $°C$ during the last century can be verified with the simulated transient temperature response.

In order to partly overcome these drawbacks, a broad range of conceivable scenarios have been generated. Simulations with these scenarios can be interpreted as a kind of uncertainty analysis. The results obtained with these scenarios are only indicative of future pathways; they do not pretend to be future predictions.

Chapter 2

General Model description of IMAGE

2.1 Introduction

The Integrated Model to Assess the Greenhouse Effect (IMAGE) is a policy oriented model based on scientific principles. More specifically, it is a parameterized simulation model, developed for the calculation of historical and future emissions of greenhouse gases on global temperature and sea level rise and ecological and socio-economic interests in specific regions. The model is based on a large variety of data derived from both an extensive study of the literature and knowledge transfer resulting from consultations with specialist experts. In this way the problem could be dealt with via a multi-disciplinary approach, combining different fields of research.

The greenhouse problem is modelled as a dynamic system which evolves in time as a non-stationary Markov chain, with discrete time steps of half a year and a simulation time of 200 years, from 1900 to 2100. The year 1900 is chosen as starting year for the simulation, symbolizing the end of the pre-industrial area. The system is split up into a number of different subsystems, which are modelled by linking sub-models, called modules. From a mathematical viewpoint, the model is a sequence of first order differential equations, and ordinary algebraic equations, which are solved using a Runge-Kutta numerical algorithm. The Runge-Kutta method is used as a default algorithm, but more sophisticated methods as Gear's Stiff have also been used to solve this series of equations.

IMAGE is a deterministic computer simulation model, which consists of interlinked modules, each of them describing a specific element of climate change. The modules are linked in a simple way: the output of one module serves as input to the next. At the highest aggregation level the framework of IMAGE consists of the following modules: a source module, an emission module, a concentration module, a climate module, a sea level rise module and a socio-economic impact module for the Netherlands, as shown in figure 2.1.

At a lower aggregation level each module is itself modular built-up, which will be described in chapter 3 and in what follows. Figure 2.2 shows the modular structure of IMAGE in greater detail. Here an arrow from one module (component) to another represents a driving influence from the first component to the second. The modular structure allows improvements to be implemented gradually without affecting the basic structure. At its inception IMAGE was set up primarily as a tool for long term greenhouse policy analysis and demonstration sessions, running on a microcomputer. The original simple parameterized model structure has been expanded with a number of elaborate modules without interfering with the original purpose. Basically IMAGE tries to capture as much as possible of the cause-effect relationship with respect to climate change. The causes are considered at the global level while, so far, impacts beyond global mean temperature and sea level rise have been included for the Netherlands only. Evidently, such a model can never be complete. New insights and additional scientific knowledge demand for continual updating, improvement and extension of the current model.

2.1. INTRODUCTION

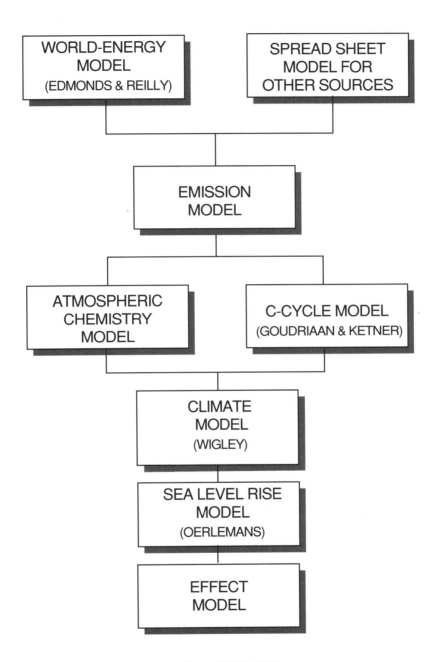

Figure 2.1: Modular construction of IMAGE.

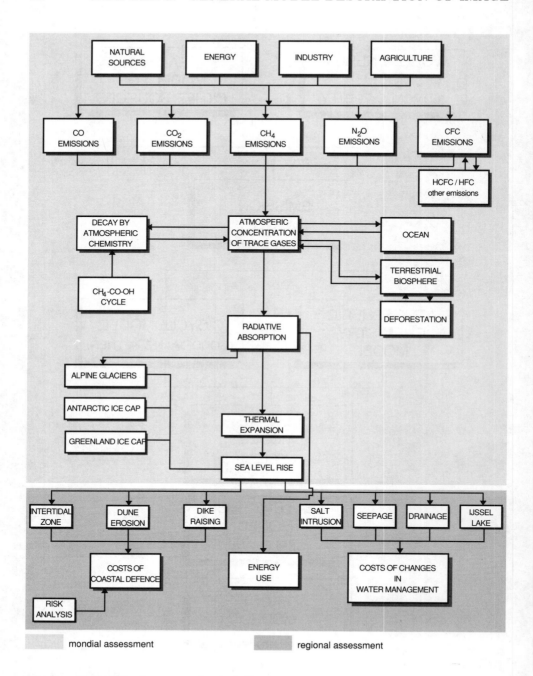

Figure 2.2: The Integrated Model to Assess the Greenhouse Effect (IMAGE).

2.2 Description of the Various Modules

2.2.1 Emission Modules

At present the model includes the trace gases CO_2, CH_4, CO, N_2O, CFC-11 and CFC-12. In the emission modules current global, annual estimates of historical emissions have been incorporated for the period 1900 to 1985. For the period 1985 to 2100 four sets of scenarios were chosen. An emission scenario is defined as a possible future development, without pretending that any probability is a prediction. Furthermore, we define a set of scenarios as a similar, consistent development for all trace gases. The underlying scenario assumptions, which will be discussed in more detail later, are based on a study of the different sources of trace gas emissions grouped as: nature, energy, agriculture and industry. At present the most detailed emission module is the energy module, for which the PC-version of the Edmonds and Reilly model is used (Edmonds and Reilly, 1986). The IEA/ORAU, long-term, global energy model was developed by Joe Edmonds and John Reilly at the Institute for Energy Analysis from 1980 onwards. The primary goal of the model is to assess long-term energy paths by considering economic, demographic, technical, and policy factors. The model is disaggregated into nine regions: USA, OECD West, OECD Asia, Centrally Planned Europe, Centrally Planned Asia, Middle East, Africa, Latin America, South and East Asia. The model is intended for making long-term projections, from 1975 to 2075, with time steps of 25 years. In the integrated IMAGE version the results of intervening years are interpolated and, for the period 2075 to 2100, results are extrapolated.

For each region the supply of and demand for energy is forecasted, as well as world and regional energy prices. Finally, CO_2 emissions are calculated by associating the release of carbon with the consumption of oil, gas, and coal. Besides carbon dioxide emissions, methane emissions from coal mining and gas distribution and nitrous oxide emissions from combustion are also related to the results of scenario calculations with this model in a simple way (by spread sheet models). The main parameters that are varied during scenario analysis are population, economic growth (labour productivity), end use and conversion efficiency, cost developments for non-fossil fuel energy sources, environmental costs and carbon taxes. So far the agricultural sources of methane and nitrous oxide have been scaled to developments in the relevant areas (e.g. nitrous oxide emissions proportional to the consumption of nitrogenous fertilizer, which is again a function of arable land development). CO_2 emissions from deforestation are not an exogenous input, but are calculated in the carbon cycle module, as will be discussed in the next section.

2.2.2 Concentration Modules

The emission modules provide the input for the concentration module. The emission and concentration of CO_2 is linked to an ocean module, and a deforestation module, together reflecting the carbon cycle. The CO_2 emission by deforestation is simulated in a separate deforestation module and integrated in the biosphere component of the carbon cycle module (Swart and Rotmans, 1989b). The C cycle module is an extended, modified version of the carbon cycle simulation model of Goudriaan and Ketner (1984). According to the present scientific opinion the assumptions with respect to the CO_2 fertilization effect have been relaxed (Swart and Rotmans, 1989b).

The structure of the greenhouse trace gas module is quite different from that of CO_2. The removal of the greenhouse gases concerned by atmospheric chemical processes is an essential feature. The spatial dimension of the simulated trace gas concentration is zero. Generally, the trace gas concentrations of CH_4, CO, N_2O, CFC-11 and CFC-12 are expressed as:

$$pX(t) = pX(t-1) + \int_{t-1}^{t} (convfX * emX(\tau) - remvlX * pX(\tau - \Delta\tau))d\tau \qquad (2.1)$$

with:
$pX(t)$ = tropospheric concentration of a trace gas at time t (in ppb)
$emX(t)$ = global emission of a trace gas at time t (in Tgy^{-1})
$convfX$ = conversion factor of trace gas X (in $ppb\ Tg^{-1}$)
$remvlX$ = removal rate of trace gas X (in y^{-1})

The concentration of methane is derived from the global CH_4-CO-OH cycle by simulating the main atmospheric chemical processes influencing the global concentrations of these trace gases. Since a large fraction of the increase of the concentration of methane in the atmosphere is most probably caused by CO competing for OH radicals, inefficient combustion in the energy sector also contributes to the greenhouse effect via this route.

The removal rates of CH_4 and CO are determined by the uptake, transport and oxidation rates of these gases, the latter being dependent on the OH concentration (Thompson and Cicerone, 1986, Khalil and Rasmussen,

2.2. DESCRIPTION OF THE VARIOUS MODULES

1985, Brühl and Crutzen, 1988, Isaksen and Høv, 1987, Logan et al., 1981, Crutzen and Graedel, 1986b, Rotmans and Eggink, 1988, Swart, 1988).

For CFCs the removal rate is supposed to be inversely proportional to the atmospheric lifetime, which is assumed to be constant (Rotmans, 1986). For CFC production, figures are input to the emission module which take account of the delay between production and emission, which is assumed to be different for different applications (Miller and Mintzer, 1986). Finally nitrous oxide concentrations are computed from emissions by taking a constant atmospheric lifetime into account.

2.2.3 Climate Module

The calculated trace gas concentrations serve as an input for the climate module. The parameterized radiative convective module, including different feedbacks, is based on Wigley (1985, 1987). The total change in radiative forcing (ΔQ_{tot}) resulting from concentration changes of CO_2, CH_4, N_2O, CFC-11 and CFC-12 is modelled according to Wigley (1987) and Ramanathan et al., (1985):

$$\begin{aligned}
\Delta Q_{tot} &= 6.23 * ln(pCO_2/pCO_2in) \\
&+ 0.0398 * (\sqrt{pCH_4} - \sqrt{pCH_4in}) \\
&+ 0.105 * (\sqrt{pN_2O} - \sqrt{pN_2Oin}) \\
&+ 0.27 * pCFC - 11 + 0.31 * pCFC - 12
\end{aligned} \qquad (2.2)$$

with:
ΔQ_{tot} = total change in radiative forcing
$pCO_2, pCH_4, pN_2O, pCFC$ = concentrations of CO_2, CH_4, N_2O and CFC
$pCO_2in, pCH_4in, pN_2Oin$ = initial concentrations (at time = 1900)

The resulting global mean equilibrium surface temperature rise can be calculated from (2.2) by dividing the radiative forcing by a climate feedback factor in which the water vapour factor is explicitly taken into account (Dickinson 1986, Wigley 1985, Tricot and Berger 1987, Ramanathan et al., 1985 and Health Council 1983). Furthermore, the transient temperature response is calculated, taking into account a time lag which suppresses the equilibrium temperature, based on Wigley and Schlesinger (1985).

2.2.4 Sea Level Rise Module

The resulting temperature changes form the input to the sea level rise module. The effects of global warming on the potential sea level are determined by five processes: thermal expansion of ocean water, melting of alpine glaciers, and ablation of the Greenland ice caps, accumulation of the Antarctic, and a natural trend:

$$\Delta Zsp(t) = \Delta Zsp_{thex}(t) + \Delta Zsp_{glac}(t) + \Delta Zsp_{Gr}(t) + \Delta Zsp_{Ant}(t) \\ + \Delta Zsp_{nat} \quad (2.3)$$

with

$\Delta Zsp(t)$	= total sea level rise at time t (in cm)
$\Delta Zsp_{thex}(t)$	= sea level rise due to thermal expansion of the ocean (in cm)
$\Delta Zsp_{glac}(t)$	= sea level rise due to melting of glaciers (in cm)
$\Delta Zsp_{Gr}(t)$	= sea level rise due to net increase of ablation of Greenland (in cm)
$\Delta Zsp_{Ant}(t)$	= sea level rise due to net increase of accumulation of Antarctica (in cm)
ΔZsp_{nat}	= natural sea level rise trend. Based on the difference between the observed sea level rise and the calculated sea level rise (in cm).

The information necessary for the description of the various complicated aspects of the phenomenon sea level rise is derived from Barnett (1983), Gornitz et al. (1982), Meier (1984), Revelle (1983), Robin (1986), United States Department of Energy (1985), Van der Veen (1986), Barth and Titus (1984), Oerlemans (1987), and Oerlemans (1989), and has been integrated and aggregated to a high level of abstraction.

The thermal expansion effect is divided into a uniform expansion for the mixed layer (0–75 m) of the ocean module of Goudriaan and Ketner (1984) and, by differential equations, a delayed expansion effect for the layers below (75–1000 m), determined by the transient temperature response calculated in the climate module. For the contributions of the glaciers, Greenland and Antarctica, differential equations have been incorporated, containing input-output factors (in $mm/\Delta T *^\circ C$).

2.2.5 Socio-Economic Impact Module

A socio-economic impact has been developed for the Netherlands, describing the consequences of an accelerated sea level rise for three kinds of coastal defence systems, namely dikes, dunes and intertidal areas (Den Elzen and Rotmans, 1988). Additionally, a risk analysis may be performed, calculating the future theoretical chance of inundation for the western and northern coastal zone of the Netherlands, compared with the present so-called "Delta norm", which involves a risk of inundation once every 10,000 years. Finally, for four elements of inland water management, the impact of sea level rise has been modelled: salt intrusion, seepage, drainage and the management of the IJssel Lake. Given four global scenarios, estimates are made of the regional costs in coastal defence and water management as a result of adaptation to the impacts of regional climate change and sea level rise. For other social sectors, such as agriculture and energy use, only tentative conclusions have been drawn.

2.3 Description of Scenarios

2.3.1 Introduction

Model calculations have been performed with four sets of scenarios, which are based on consistent assumptions for each trace gas. A general survey of these assumptions is given in Table 2.1. The sets of scenarios are meant to encompass the possible global, socio-economic developments in order to illustrate the impact of different future pathways on the greenhouse effect. The highest scenario, A: *unrestricted* trends, assumes a continuation of economic growth, not limited by environmental constraints. Scenario B: *reduced* trends, is meant to include the implementation of environmental measures presently being considered to control other environmental problems, such as acidification and eutrophication, having important side-effects for the greenhouse effect. Scenario C: *changed* trends, assumes the enforcement of a stricter environmental control, at least partly influenced by international concern about the greenhouse effect as expressed at some recent conferences (see Jäger 1988, Environment Canada 1988). Finally, scenario D: *forced* trends, assesses the possibilities of maximum efforts towards global sustainable development. World population growth, a factor that is believed not to be influenced by greenhouse policies, is assumed to approach 10.8 billion in 2100 in all scenarios.

2.3.2 Energy Supply

The combustion of fossil fuel is the most important human activity contributing to the emissions of greenhouse gases. The emissions of CO_2 can be quantified reasonably well for different energy scenarios. In the literature distinctions between scenarios are usually very similar to those described above. The PC-version of the IEA/ORAU Long-Term Global Energy CO_2 Model (Edmonds and Reilly 1985, 1986) has been fully integrated into IMAGE. In this work four Energy CO_2 scenarios have been reproduced using this model, using the input data given by Mintzer (1987). His scenario assumptions coincide very well with the four types of world development described above. Like population growth, labor productivity growth — the main driving factor for economic growth — is assumed to be common for all scenarios and, while varying among the different regions, it averages 1.8 % annually. The energy related emissions of CH_4 and N_2O are less well-defined than those of CO_2 but may play an important role in the greenhouse problem. For the purpose of the IMAGE calculations average estimates for the emissions of the different sources of these gases are assumed to be valid for 1985 and future emissions are determined by assessing developments of these sources along the lines described above. Unabated CH_4 emissions from the mining of coal and the exploitation of natural gas are assumed to be proportional to the consumption of these fuels. In particular, because of the expected long term increase in the use of coal, better information is needed about the methane emission rates from coal mining (both open pit and deep mining) and the nitrous oxide emission rates from coal combustion.

The most important policy tools are efficiency improvements for supply, conversion and end use, cost assumptions for renewables and synfuels, environmental costs (or taxes) for supply and end use and abatement of energy related emissions of methane and nitrous oxide.

In the high scenario (A: *unrestricted trends*) economic growth is based on an increase in fossil fuel consumption, especially coal because of the higher prices of the decreasing oil and gas resources. It should be noted that in the results from the Edmonds and Reilly model recent discoveries of major natural gas reserves have not yet been included. In this scenario environmental concerns neither alter ways of life nor lead to substantial efforts to reduce emissions. Introduction of renewable energy is retarded (e.g. solar energy at US$ 20.-/GJ) and production of synfuels is enabled by relatively low prices (e.g. non-energy prices for synoil at US$ 3.50/GJ, for syngas at US$ 2.75/GJ).

2.3. DESCRIPTION OF SCENARIOS

TRACE GAS → SCENARIO ↓	CO_2	CH_4
Continued trends A	- major increase energy consumption based on fossil fuels (coal) - no incentives for increase of energy efficiency - increasing per capita energy use (espec. in third world) - no simulation renewable and other non-fossil energy - growth not limited by environmental considerations - rapid conversion of available forest resources: shifting cultivation, fuel wood, commercial wood, cattle breeding	- energy scenario see CO_2 major increase in CH_4 emissions from coal mining and natural gas losses - increasing number of cattle (developing countries) - increasing area rice paddies - continuing conversion of tropical forests (emissions from biomass burning, increasing number of termites) - average estimate anthropogenic sources increasing emissions from waste dumps, especially from growing 3rd world population
Reduced trends B	- mainly because of price increases (scarcities, some environmental costs): - slight shift towards non-fossil energy sources - minor increase per capita energy consumption - increasing energy efficiency - gradual reduction deforestation rates	- energy scenario see CO_2: important emissions from coal mining minor increase number of cattle (developing countries) - reduction deforestation rates - increasing use of gas from wastes/manure - average estimate anthropogenic sources
Changing trends C	- environmental concerns lead to modest policy changes: - incentives for increase in energy efficiency - incentives for shifts towards renewable energy sources - global efforts to reduce deforestation rates	- energy scenario see CO_2 minor increase energy related CH_4 emissions - after 2000 stabilization number of cattle - increase rice production from intensification, not expansion of wet area growing recovery of CH_4 from wastes, coal mining - reduction deforestation - low estimate anthropogenic sources
Forced trends D	- major policy changes because of greenhouse warming concerns: - reduction use fossil fuels - strong efforts towards greater energy efficiency - stagnation world economy - major effort halting deforestation by forest management and reforestation programs	- energy scenario see CO_2 decreasing coal use - major recovery CH_4 from wastes and other sources - halting deforestation - limitation losses from leakages natural gas - after 2000: reduction number of cattle (limitation consumption meat and dairy) - reduction CO emission - low estimate anthropogenic sources

Table 2.1: Policy assumptions in scenario sets.

TRACE GAS → SCENARIO ↓	N_2O	CFCs
Continued trends A	- energy scenario see CO_2: enormous increase N_2O emission from coal use - no emission control measures - rapid increase fertilizer use in third world up to European levels - average estimate anthropogenic sources - continuing deforestation (emission from burning and possible major influence soil emissions)	- continuing trends; continuing growth of economy and cfc-consumption; no implementation UNEP-ozone-protocol - no recycling
Reduced trends B	- energy scenario see CO_2: large increase N_2O emissions from coal consumption - gradual increase fertilizer use in third world - gradual reduction deforestation	- implementation basic ozone protocol by ratifiers - production increase by non-ratifiers - minor reduction CFC losses during production/use - no increase in applications - total result: stabilization of production
Changing trends C	- energy scenario see CO_2: increase N_2O emissions from coal consumption - slight reduction emission rates from fossil fuel combustion - slow increase fertilizer use in third world - major global effort towards halting deforestation	- gradual upgrading ozone protocol: greater production decrease for all countries - reduction of losses and stimulation of recycling - total result: gradual decrease of production
Forced trends D	- energy scenario see CO_2: decreasing N_2O emission due to shift towards non-fossil energy sources and implementation of N_2O emission control technologies - global program to ontroduce proper forest management, stop deforestation and start large scale reforestation (influence N_2O emissions for the future unclear) - limited growth fertilizer use - controled growth of fertilizer use: type and method of application: limiting N-losses (through denitrification)	- strongly upgraded protocol - stagnancy world economy - elimination of losses - major recycling efforts cfc refrigerants, foam applications

Table 2.1: (continued)

2.3. DESCRIPTION OF SCENARIOS

End use efficiency increases by only 0.2% annually and the efficiency of coal supply ($0.75\%yr^{-1}$) increases more than the efficiency of fuels with lower CO_2 emission factors (e.g. $0.3\%yr^{-1}$). No major environmental costs are assumed. Since coal has the highest emission rates this scenario leads to high CO_2 emissions. The effects of these high emissions may even cause a negative feedback in the next century, climatological change affecting economic growth and food production. This has not yet been taken into account. High temperature combustion of coal and other fossil fuels causes increasing N_2O emissions (Hao et al. 1987, Kavanaugh 1987). The release of methane during the exploitation of coal reserves may cause a sharp increase in CH_4 emissions. However, further research into this aspect for different methods of coal mining is needed to confirm this statement. This is particularly interesting, since recent research suggests that the contribution of fossil sources to methane increase may have been underestimated (Lowe et.al., 1988). Because of a rather stable supply of natural gas in all scenarios in absolute terms, CH_4 emissions from leakages will not alter much. It may be assumed that efforts to reduce leakages will compensate for emissions caused by possibly increasing transport distances.

In the *reduced trends* scenario (B) the introduction of non-fossil fuels is accelerated and energy efficiency increased because of higher prices of fossil fuels, influenced by scarcity and environmental measures. The costs of solar energy are assumed to be lower than in scenario A (US$16.50/GJ) and end use efficiency increases by 0.8% annually. Supply efficiency for gas and oil are assumed to increase faster (0.3% annually) than for coal (0.2%). Synfuels are relatively more expensive (US$ 4.25 non-energy costs for synoil and US$ 3.15 for syngas). Moderate environmental costs for coal are introduced. Methane is a useful gas and therefore the reduction of losses is not only interesting from the environmental point of view. A gradual increase of the methane recovery from coal mining up to 25% in 2100 is assumed. This might be a scenario including the implementation of the environmental strategies presently being considered. CO_2 and CH_4 emissions from coal combustion and mining are still high.

In the *changed trends* scenario (C) environmental strategies will be upgraded. Incentives accelerate the introduction of renewable energy sources (e.g. solar US$ 15.00/GJ). End use energy efficiency increases by 1% annually and supply efficiency increases more for gas ($0.4\%yr^{-1}$) than for oil (0.3%). Higher prices delay the introduction of synfuels (synoil US$ 5.- and syngas US$ 4.- non-energy prices). Again, 15% of the methane from coal mining is assumed to be recovered in 2100. Higher environmental costs are introduced for coal. Increased energy consumption, especially in the third world, will still cause a considerable increase of coal consumption.

The only scenario showing a continuation of or, in the long run, a return to present day emission rates is the *forced trends* scenario (D). Concern about the global environment will lead to a change in lifestyle in the developed world, combined with the introduction of very energy efficient technologies using renewable energy sources both in developed and developing countries. This will lead to low per capita energy consumption of fossil fuels rather than the introduction of expensive emission control technologies, enabling a sustainable development of global societies (Mintzer 1987, Goldenberg et al. 1987a and 1987b, Cheng et al. 1986). Costs of solar energy are assumed to arrive at US$ 12.-/GJ in 25 years. End use efficiency increases by 1.5% annually and supply efficiency for natural gas ($1.5\% yr^{-1}$) increases much faster than for oil (0.6%) and coal (0.2%). High prices (synoil US$ 7.- and syngas US$ 5.50 non-energy costs) and low energy consumption prevent the introduction of synfuels. Environmental costs (taxes) proportional to their carbon content are applied to supplies of the different types of fuel and their consumption. Remaining N_2O emissions are assumed to be reduced by technological measures. Half of the CH_4 emissions from coal mining are finally recovered.

2.3.3 Agriculture

Although energy production and consumption are the main contributors to the greenhouse effect, agricultural activities also influence the emission of trace gases: methane from cattle and rice paddies, nitrous oxide from fertilization and probably carbon dioxide from deforestation. Again the preliminary scenarios for methane and nitrous oxide (both anthropogenic and natural) are based on average base year estimates from a variety of sources, among which are Sheppard et al. (1982), Khalil and Rasmussen (1982, 1984, and 1985), Atmospheric Ozone (1985), Van Ham (1987), Bolle et al. (1986), Holzapfel-Pschorn and Seiler (1986), Bingemer and Crutzen (1987), Crutzen et al. (1986a), Seiler (1984), and Bartlett et al. (1985) for methane, and Keller et al. (1987), Van Ham (1987), Conrad et al. (1980, 1983), Marland and Rotty (1985), Hignett (1985), and Anderson and Levine (1987) for nitrous oxide. Scenario assumptions are rough estimates; in comparison to the figures of the Food and Agricultural Organization (1987) they tend to be low.

In the *unrestricted trends* scenario it is assumed that food production

2.3. DESCRIPTION OF SCENARIOS

will keep up with population growth through massive intensification of agricultural production by way of fuel or mined mineral-based technologies, including fertilizers and irrigation. The sustainability of such a scenario may be questioned.

Because of an increase of agricultural land and the intensity of fertilizer application the use of nitrogenous fertilizers will increase the still uncertain N_2O emissions by more than fourfold by the year 2100. For predicting the increase in arable land logistic curves are used, fitted to historical FAO data and consistent with estimates of potential arable land. Although the area available for rice cultivation will probably not be extended very much, since the remaining arable area will become scarce and erosion will cause loss of fertile soil, irrigation of the present rainfed area has been assumed to lead to increases of CH_4 emissions by almost 80% in 2100. Again logistic curves are applied utilizing FAO data for historical development of rice paddy area. With an increasing number of countries reaching higher levels of prosperity, an increasing consumption of meat and dairy products will lead almost to a doubling of the number of methane producing cattle, mainly in the now developing world.

The unrestricted economic growth in this scenario will further exploit the tropical forests. Cattle breeding, cultivation of export products, production of logs and fuelwood, mining or infrastructural projects, all force the increasing rural population further into the remaining forests, practising a form of slash and burn agriculture, which is often not sustainable due to shorter rotation times and unsuitable soils (World Resources Institute 1986, 1987, Molofsky et al. 1986). This affects the emissions of several trace gases: more CO_2 will be released into the atmosphere. The effects as to CH_4 and N_2O are still unclear. Tropical soils serve as a sink for CH_4, higher rates being observed for savannahs than for forest soils. N_2O emissions from tropical forest soils are important and may increase after forest conversion. As the number of termites increases when tropical forests are being converted, emissions by termites may increase.

In the *reduced trends* scenario (B) the high costs of agricultural inputs to increase productivity and irrigated areas decrease the present growth rates of fertilizer use and irrigation, thus limiting the agricultural emissions of CH_4 and N_2O. The number of cattle gradually increases by 75% in 2100, fertilizer use increases by 250% and rice paddies by about 40%. Deforestation rates are gradually slowing down.

In the *changed trends* scenario (C) the changes of scenario B are accelerated: increases in productivity are sought in intensification based on agro-ecological practises rather than technological means, although use of fertilizer remains essential in order to meet food demands, causing a 150% increase in fertilizer consumption globally in 2100. A global effort to control deforestation is initiated. The number of cattle increases by only 50% compared to a doubling of the human population. Agricultural wastes are increasingly used for energy production.

The *forced trends* scenario (D) reflects a worldwide change of attitude towards agricultural practices. The use of fertilizers and pesticides is limited to the necessary minimum agricultural development aiming at full sustainability at the local or regional level. This scenario is dependent on the spread of sustainable agro-ecological techniques, as yet to be fully developed (e.g. see Dover and Talbot 1987). Yet a global increase in nitrogen fertilizer (80% in 2100) is still assumed to be unavoidable. In order to optimize food efficiency meat consumption is limited, leading to only a small increase (10% in 2100) in the number of cattle and the associated methane emissions.

All agricultural wastes (crop residues, manure) are used for fertilizers or energy production. An effective global program stops deforestation and stimulates proper forest management and reforestation. All these activities limit emissions of CO_2, CH_4 and N_2O, and even change the direction of present trends.

2.3.4 CFC Use

In the *unrestricted trends* scenario it is assumed that the ozone protocol is not fully implemented: production and use will continue to increase, albeit at a moderate rate, and will stabilize in the first half of the 21st century.

In the *reduced trends* scenario the Montreal protocol (United Nations Environment Program 1987) is supposed to be implemented by linearly decreasing the production towards the limits set for 1993 (80%) and 1998 (50%), making use of the allowance (10% and 15% respectively) to 'satisfy basic domestic needs' and 'for the purpose of industrial rationalization' followed by stabilization after 1999. The allowed growth of production planned in the protocol is not taken into account because of lack of data, while no substances other than CFC-11 and CFC-12 are yet being considered in IMAGE. The objectives are reached by turning to substitutes, limiting losses and gradually increasing recycling.

In the *changed trends* scenario the basic protocol is upgraded to a limit of 15% of the 1986 production and consumption figures for 1998, after which annual production and consumption are assumed to stabilize. No additional production capacity is built, since this seems to be useless under these circumstances. The efforts towards the development of 'safe' substances, recycling and loss reduction are increased.

In the *forced trends* scenario a total ban is assumed in 2050 in addition to the 85% reduction in scenario C. A possibly remaining use for purposes that are considered to be essential is neglected.

2.3.5 Other Trends

Methane emissions from waste dumps are increasing with population growth and the increasing level of prosperity. In the scenarios from A to D an increasing percentage of methane recovery from waste is assumed (2100 0%, 20%, 30% and 50% respectively).

Other factors that may possibly have an influence are kept constant because of lack of even qualitative information. Forest and bush fires may increase with population, and the increasing CO_2 concentrations will probably increase the release of CH_4 emissions from an aerobic microbiological sources. The reclamation of wetlands may reduce CH_4 emissions, but temperature rises may melt the permafrost in the Arctic regions, increasing the 'wetland' areas.

2.4 Model Deficiencies and Future Developments

The simulation model is a reflection of the current state of knowledge. Current knowledge being far from complete, the model has structural limitations. An important spin-off of IMAGE is the possibility of recognizing gaps in our knowledge, which is necessary for a better prediction of the effects of the enhanced greenhouse phenomenon. This may lead to the allocation of future research priorities.

Ozone and ozone depleting substances other than CFC-11 and CFC-12 are planned to be added to IMAGE in the near future. Secondly, an integrated energy module will be developed, based on the LEAP model of the Energy and Systems Research Group in Boston (ESRG, 1988). Thirdly, an energy balance module is being developed, including the lag effects between the equilibrium and transient temperature rise, and extended by a pattern of regional differentiation. A number of internal feedback processes will be incorporated (especially biogeochemical feedbacks such as changes in methane emissions, ocean CO_2 uptake, and vegetation albedo). Furthermore the atmospheric chemistry within IMAGE will be made more complete, with the addition of NO_x and O_3. On a European scale the ecological effects of climate change will be added to IMAGE in the near future.

In consequence of the modular structure of IMAGE new information can generally be incorporated without affecting the model structure.

2.5 Comparison with Other Models

Recently in Washington IMAGE was compared to the Model of Warming Commitment (MWC) of the World Resource Institute (Mintzer, 1987), and the Atmospheric Stabilization Framework (ASF) of the EPA (EPA, 1989), which are the only comparable integrated greenhouse policy models published to date. Generally the model builders concluded that the frameworks of these models were very similar. However comparing IMAGE with ASF (EPA model) the following major differences should be mentioned (Response Strategies Working Group, 1989a and 1989b):

- o The structures differ in aggregation level: the EPA model is more detailed than IMAGE, especially with respect to the energy end use, the industrial (CFCs), and agricultural sectors;

- o The ASF atmospheric chemistry is more complete;

- o The ASF includes emissions of more trace gases than does IMAGE (i.e. ozone-depleting substances other than CFC-11 and CFC-12, NO_x, and ozone as a greenhouse gas);

- o IMAGE includes a fully integrated carbon cycle, while ASF uses exogenous variables for the carbon cycle and biogenic emissions;

- o IMAGE includes emissions of non-methane hydrocarbons, which are not included in ASF;

- o IMAGE covers a timespan between 1900 to 2100, while ASF simulation time is from 1985 to 2100;

- o IMAGE includes a number of global (sea level rise) and national impact modules, which are not included in ASF;

- o IMAGE has a more flexible structure, allowing alternatives to be implemented easily.

Comparing IMAGE with MWC (WRI model) the following similarities and differences appear:

o IMAGE and MWC have a comparable aggregation level;

o The present energy end use and CFC modules of MWC are more detailed than in IMAGE;

o IMAGE includes a carbon cycle, while MWC uses exogenous variables instead of a carbon cycle;

o CO_2 emission by deforestation is in IMAGE not an exogenous variable like MWC, but is integrated in the biosphere component of the carbon cycle module;

o For methane, IMAGE includes a CH_4-CO-OH cycle incorporating emissions of methane and carbon monoxide, while MWC only takes CH_4 concentrations into account;

o The most obvious additional features of IMAGE are the effect modules (for instance the sea level rise module)

IMAGE runs were performed with the same input emission scenarios as used in the EPA model and the WRI model. It appeared that the resulting equivalent CO_2 concentrations and temperature changes were not significantly different (Response Strategies Working Group, 1989a, 1989b). Among other things this exercise confirmed the reproducibility of the results. One important reason for the differences appeared to be the different treatment of the biospheric component of the carbon cycle.

2.6 Discussion

Over the last few years the utility of an integrated methodology for evaluating different futures with respect to climate change has been proved extensively. IMAGE gives clear, long-term relationships between causes and effects of the enhanced greenhouse effect, based on the latest scientific evidence, but without requiring detailed scientific knowledge from the user. Given that the Edmonds and Reilly model has 25 year time steps, IMAGE is less suitable for the assessment of details of the effects, feasibility and costs of different policy options in the short run. Attention will be given to this drawback over the following years in collaboration with other groups.

Chapter 3

The Carbon Cycle Model

3.1 Introduction

Carbon dioxide (CO_2) occurs naturally in the atmosphere, and plays an important role in almost all living mechanisms. The natural carbon cycle encompasses exchange of CO_2 between the atmosphere, oceans and the terrestrial biosphere of hundreds of billions tons of carbon a year. Compared with these tremendous quantities the extra man-made addition through the burning of fossil fuels and changing land use is only a small contribution to the carbon cycle. Nevertheless this minor anthropogenic injection is supposed to account for the imbalance of the carbon cycle and thus for the increase in the CO_2 concentration. However, only about 40% of the man-made CO_2 emissions remains in the atmosphere. The remainder is taken up by the oceans and terrestrial ecosystems. The atmospheric CO_2 level has increased approximately 25% from 1900 to 1989; additionally, since 1958, the starting year of systematic accurate measurements of atmospheric CO_2 at Mauna Loa, the CO_2 level has increased by more than 11%. The total anthropogenic CO_2 flux to the atmosphere is surrounded by uncertainties (USDOE, 1985a). In fact only the resulting flux from the combustion of fossil fuels is well known, in contrast to the flux from the terrestrial biota to the atmosphere or vice versa, which is still debated. Some argue that the terrestrial biosphere is a net sink of atmospheric CO_2 (Lugo and Brown, 1980, Goudriaan and Ketner, 1984, Esser, 1987, Harvey, 1989b), while others argue that it is a net source (Houghton et al., 1983).

IMAGE contains a carbon cycle model which is merely an interpretation of the complex global carbon cycle mechanism and which is described in detail below.

3.2 Model Description

A carbon cycle model has been developed which consists of several linked modules: an emission module, an atmospheric concentration module, an ocean module, a terrestrial biota module and a deforestation module. The CO_2 emission module provides the input for the concentration module of CO_2. Both modules are linked to an ocean module and a terrestrial biota module, together reflecting the carbon cycle. The latter is a modified version of the carbon cycle model of Goudriaan and Ketner (1984). All C cycle models assume that the atmospheric-ocean-terrestrial biosphere system was in steady-state prior to the industrial era (Kohlmaier et al., 1981, Houghton et al., 1983, Emanuel et al., 1984, Esser, 1987), although this is questioned by Lugo and Brown (1986).

In consequence of the coupling of the emission, concentration, ocean and terrestrial biota modules the airborne fraction has been defined in a specific sense: the airborne fraction is the fraction of the total CO_2 emissions from fossil fuels and biosphere changes that remains in the atmosphere. In formula form:

$$\begin{aligned} Af &= \frac{\text{net increase of } CO_2 \text{ in the atmosphere}}{\text{fossil fuel emission } - \text{ net terrestrial biota uptake}} \\ &= \frac{\text{fsem} - \Delta \text{ ocean} + \text{bioem} - \text{bioupt}}{\text{fsem} + \text{bioem} - \text{bioupt}} \\ &= \frac{\text{fsem} - \Delta \text{ ocean} - \Delta \text{ biota}}{\text{fsem} - \Delta \text{ biota}} \end{aligned} \quad (3.1)$$

with:
Af = airborne fraction
fsem = fossil fuel emission (in GtC)
Δocean = CO_2 uptake by the ocean (in GtC)
Δbiota = net biota uptake = bioupt - bioem (in GtC)
bioem = emission released from terrestrial biota by human disturbance (in GtC)
bioupt = uptake by terrestrial biota, equivalent with the total net ecosystem production (in GtC)

3.3 Emissions Module

There are three anthropogenic sources of CO_2 emission: 1. Fossil fuel combustion, 2. Production of cement, 3. Land use change.

3.3. EMISSIONS MODULE

Only the first two issues will be treated in this section, the third one will be dealt with in the section about the deforestation module. Figures concerning the contribution of fossil fuels to CO_2 emission are known from 1860 (Watts, 1982). Marland and Rotty (1984) and later on Rotty and Masters (1985), and Rotty (1987) have adjusted these figures, although the differences are small. Historical fossil fuel emissions figures have been put into IMAGE for the period 1900–1985, involving an uncertainty range of about 5%.

For the period before 1950 no cement production data are on record, whereas the cement production from 1950 is reasonably well known. For the period 1950–1985 cement figures, based on Marland and Rotty (1984) have been implemented.

For the period 1985–2100 four different Energy CO_2 scenarios have been introduced (Rotmans et al., 1990a). These scenarios are generated with the PC-version of the IEA/ORAU Long-Term Global Energy CO_2 Model (Edmonds and Reilly, 1986), which has now been fully integrated into IMAGE. The IEA/ORAU Long-Term Global Energy CO_2 Model was developed by Jae Edmonds and John Reilly to run on a personal computer and was released in 1985. The model calculates alternative long-term energy strategies and the effects on global economic activity and finally estimates the emissions of trace gases resulting from the production, transportation, distribution, and consumption of fossil fuels.

In the *unrestricted trends* (A) scenario economic growth is based on an increase in fossil fuel consumption, especially coal, because of the higher prices of decreasing oil and gas resources. Environmental concerns neither alter ways of life nor lead to substantial efforts to reduce emissions. Introduction of renewable energy is retarded and production of synfuels is enabled by relatively low prices. No major environmental costs are assumed. Since coal has the highest CO_2 emission rates this scenario leads to high CO_2 emissions.

In the *reduced trends* scenario (B) the introduction of non-fossil fuels is accelerated and energy efficiency increased because of higher price of fossil fuels, influenced by scarcity and environmental measures. Moderate environmental costs for coal supply are introduced.

In the *changed trends* scenario (C) environmental strategies will be upgraded. Incentives accelerate the introduction of renewable energy sources and provides an increase in energy efficiency.

The only scenario showing a continuation of or, in the long run, a return to present day emission rates is the *forced trends* scenario (D). Concern about the global environment will lead to a change in lifestyle in the developed world, combined with the introduction of very energy efficient technologies using renewable energy sources both in developed and developing countries. Environmental costs (taxes) proportional to their carbon content are applied to supply of the different fuel types and their consumption. Figure 3.1 shows the emissions of CO_2 for the four scenarios described above.

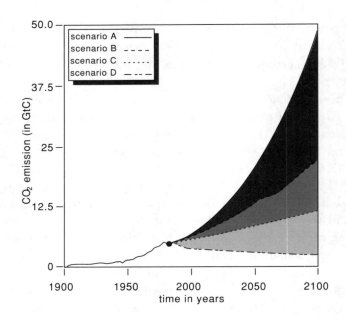

Figure 3.1: Emissions of CO_2.

3.4 Atmospheric Concentrations Module

The atmosphere is represented as a well-mixed reservoir, with a mixing time of 1 year, although in reality the atmospheric concentration of CO_2 is not uniformly mixed. The atmospheric CO_2 concentration is determined by the fossil fuel combustion, uptake of CO_2 by the oceans, flux of CO_2 from the terrestrial biota and the net ecosystem production flux. The driving forces of this system are the fossil fuel combustion and the changing land use, perturbing the global carbon cycle, and causing an imbalance in the CO_2 uptake by the oceans, and in the net ecosystem production. The atmospheric

3.5. OCEAN MODULE

CO_2 concentration is then modelled according to the following equation:

$$pCO_2(t) = pCO_2(t-1) + \int_{t-1}^{t} [ATMCF * (FSEM(\tau) + OCEA(\tau)$$
$$-TNEP(\tau) + THDIST(\tau))]d\tau \qquad (3.2)$$

with:
$pCO_2(t)$ = atmospheric CO_2 concentration at time t (in ppm)
$pCO_2(0)$ = initial CO_2 concentration at time $t = 0$, in 1900; (in ppm) is 285 ppm, varies from 270–300 ppm (Rotmans, 1986)
$ATMCF$ = factor that converts emissions of CO_2 into concentrations; is 0.471 ppm/GtC according to Brewer (1983) (in ppm/GtC)
$FSEM(\tau)$ = fossil fuel combustion flux at time t (in GtC/yr)
$OCEA(\tau)$ = flux from oceanic mixed layers to the atmosphere (in GtC/yr)
$TNEP(\tau)$ = carbon flux by total net ecosystem production (in GtC/yr)
$THDIST(\tau)$ = total carbon flux of CO_2 due to human disturbance

3.5 Ocean Module

The ocean module is a slightly modified version of that given in Goudriaan and Ketner (1984), based on De Haan (1989). It belongs to the class of ocean models which are modified versions of the basic box-diffusion model of Oeschger et al. (1975), of which the model by Björkström (1979) was the first. Emanuel et al. (1985) surveys the different types of ocean models.

In this module the ocean is divided into 12 layers. Two surface layers, a warm mixed layer of 75 meters and a cold mixed layer of 400 meters, one intermediate layer of 1005 meters and nine layers of 680 meters, represented in Figure 3.2.

Three primary processes are defined, determining the carbon concentration in the ocean: carbon transport by mass flow of water, by turbulent mixing and by gravitation. Refinements that are not yet included in the IMAGE ocean module, but are being developed in the new ocean module to increase its realism, are polar outcrop, nutrient cycling, upwelling and separate boxes for the Atlantic, the Pacific and the Indian ocean. One-dimensional models are available in which most of the refinements cited above are already incorporated (Hoffert et al., 1981, Peng et al., 1983, Bolin et al., 1983). In the ocean model of Goudriaan and Ketner (1984) marine photosynthesis was added as a fixed driving force (Goudriaan, 1988b).

It should be noted that although the one-dimensional box-diffusion models are useful for general analysis of the role of the oceans in the global carbon cycle, none of these one-dimensional box-diffusion models gives a realistic representation of the dynamics of the ocean. A realistic reflection of the ocean circulation must account for geometry. However, at present, a realistic, general circulation model of the ocean does not exist (Emanuel et al., 1985).

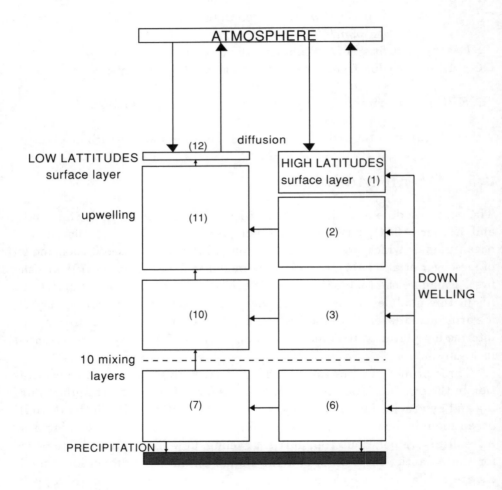

Figure 3.2: Schematic of the IMAGE 12-layer ocean model.

3.5. OCEAN MODULE

Two of the prominent processes, mass flow and turbulent mixing, can be represented in a discrete way (De Haan, 1989):

Mass flow:
$$\frac{C^{t+1}(i) - C^t(i)}{\Delta t} = u^t(i) * \frac{C^t(i) - C^t(i-1)}{\Delta D} \tag{3.3}$$

with:
$C^t(i)$ = carbon concentration in ocean layer i at time t (in tC/m^3)
Δt = time interval t (in yr)
ΔD = depth of an ocean layer (in m)
$u^t(i)$ = constant in ocean layer i at time t (in m/yr)

Turbulent mixing:
$$\frac{C^{t+1}(i) - C^t(i)}{\Delta t} = DIFF * \frac{C^t(i+1) - 2*C^t(i) + C^t(i-1)}{\Delta D^2} \tag{3.4}$$

with:
ΔD^2 = square of the ocean layer depth (in m^2)
$DIFF$ = diffusion coefficient (see below) (in m^2/yr)

These two processes, mathematically defined above, are incorporated in the following equations, together with the process of precipitation of organic material:

$$\frac{dO(1)}{dt} = \underbrace{[MFL * (C(12) - C(1))]}_{MASSFLOW}$$
$$+ \underbrace{\left[DIFF * \frac{V(1)}{D(1)} * \frac{(C(2) - C(1))]}{(D(2) + D(1))/2}\right]}_{TURBULENT\ MIXING} - \underbrace{[PRC]}_{PRECIPITATION} \tag{3.5}$$

$$\frac{dO(i)}{dt} = [1/5 * MFL * (C(1) - C(I))]$$
$$+ \left[DIFF * \frac{V(I)}{D(I)} * \frac{C(i+1) - C(i)}{D(i)}\right.$$
$$\left. - \frac{C(i) - C(i-1)}{(D(i) + D(i-1))/2}\right] - [PRC] \quad i = 2, \ldots, 5 \tag{3.6}$$

$$\frac{dO(6)}{dt} = [1/5 * MFL * (C(1) - C(6))]$$
$$+ \left[DIFF * \frac{V(6)}{D(6)} * \frac{(C(5) - C(6))}{D(6) + D(5))/2} \right] - [PRC] \quad (3.7)$$

$$\frac{dO(7)}{dt} = [1/5 * MFL * (C(6) - C(7))]$$
$$+ \left[DIFF * \frac{V(7)}{D(7)} * \frac{(C(8) - C(7))}{(D(8) + D(7))/2} \right] - [PRC] \quad (3.8)$$

$$\frac{dO(i)}{dt} = [1/5 * MFL * (C(13 - I) + ((I - 7) * C(I - 1))$$
$$- ((I - 6) * C(I)))]$$
$$+ \left[DIFF * \frac{V(I)}{D(I)} * \frac{(C(i+1) - C(i))}{(D(i) + D(i+1))/2} \right.$$
$$\left. - \frac{(C(i) - C(i-1))}{D(i)} \right] - [PRC] \quad i = 8, ...11 \quad (3.9)$$

$$\frac{dO(12)}{dt} = [MFL * (C(11) - C(12))]$$
$$+ \left[DIFF * \frac{V(12)}{D(12)} * \frac{(C(12) - C(11))}{(D(12) + D(11))/2} \right] - [PRC] \quad (3.10)$$

with:

$\frac{dO(i)}{dt}$ = change in amount of carbon in ocean layer i (in GtC/yr)
$C(i)$ = carbon concentration in ocean layer i (in GtC/m^3)
$D(i)$ = thickness of ocean layer i (in m)
$V(i)$ = volume of ocean layer i (in m^3)
$DIFF$ = diffusion coefficient ; is 4000 m^2/yr, varies in the literature from about 3700 to 6000 m^2/yr (Hoffman, 1984)
MFL = mass flow of water ; is $2.3 * 10^{15} m^3/yr$,
PRC = precipitation flux; is constant according to Goudriaan and Ketner (1984) (in GtC/yr)

The CO_2 exchange between the surface layers of the ocean and the atmosphere is buffered. The buffer factor is widely used to represent the equilibrium distribution of CO_2 between the ocean and the atmosphere. This buffer factor is dependent on the the atmospheric CO_2 concentration and the inorganic carbon content in the surface layers of the ocean. Additionally the buffer factor increases with decreasing temperature (Bolin, 1986).

3.5. OCEAN MODULE

The following equation reflects the equilibrium phase between CO_2 in the atmosphere and in the ocean (Mook and Engelsman, 1983):

$$BUFF * \frac{dO_{eq}}{O_{eq}} = \frac{dpCO_2}{pCO_2} \qquad (3.11)$$

with:
- $BUFF$ = buffer factor
- dO_{eq} = change in the amount of inorganic carbon in the ocean surface layers, in equilibrium with the atmospheric CO_2 concentration (in GtC)
- O_{eq} = amount of inorganic carbon in the ocean surface layers, in equilibrium with the atmospheric CO_2 concentration (in GtC)
- $dpCO_2$ = change in atmospheric concentration of CO_2 (in ppm)
- pCO_2 = atmospheric CO_2 concentration (in ppm)

In modelling the buffer factor the dependence on the inorganic carbon and on temperature is ignored. Only the functional relationship with the atmospheric CO_2 relation has been taken into account (Rotmans, 1986, and Health Council, 1983):

$$BUFF(t) = 4.05 * Ln(0.033 * pCO_2(t)) \qquad (3.12)$$

with:
- $BUFF(t)$ = buffer factor at time t
- $pCO_2(t)$ = atmospheric CO_2 concentration at time t (in ppm)

In this way the diffusion fluxes between the ocean and the atmosphere can be calculated:

$$FLOC1 = (O_1 - O_{1eq})/RT1A \qquad (3.13)$$
$$FLOC2 = (O_2 - O_{2eq})/RT2A \qquad (3.14)$$

with:
- $FLOC1$ = flux from the thin surface ocean mixed layer to the atmosphere (in GtC/yr)
- $FLOC2$ = flux from the higher latitude surface ocean layer to the atmosphere (in GtC/yr)
- O_1 = amount of carbon in the thin surface ocean mixed layer (in GtC)

O_{1eq} = amount of carbon in the thin surface ocean mixed layer in equilibrium with atmospheric CO_2 (in GtC)

O_2 = amount of carbon in the higher latitude ocean mixed layer (in GtC)

O_{2eq} = amount of carbon in the higher latitude ocean mixed layer in equilibrium with atmospheric CO_2 (in GtC)

$RT1A$ = residence time of carbon in the thin surface ocean layer; is 1 year according to Goudriaan and Ketner (1984)

$RT2A$ = residence time of carbon in the higher latitude surface ocean layer; is 20 year according to Goudriaan and Ketner (1984)

The net flux from the mixed ocean layers to the atmosphere is formed by the addition of FLOC1 and FLOC2, and is used in Equation (3.2), where it is written as OCEA.

3.6 Terrestrial Biosphere Module

Various models have been developed to estimate the role of terrestrial biota in the carbon cycle. Wiersum and Ketner (1989) describe three kinds of models: bookkeeping models, evaluating amounts of CO_2 release by land-use change (Avenhaus and Hartmann, 1975, Detwiler and Hall, 1988, Houghton et al., 1987); secondly, dynamic global carbon cycle models (Goudriaan and Ketner, 1984, and Esser, 1987); and finally, geochemical models, tracing historic CO_2 fossil fuel emissions in the atmospheric CO_2 budget (Peng et al., 1983, and Emanuel et al., 1984).

Although the various model estimates of the terrestrial biospheric contribution vary from a net CO_2 uptake to a net CO_2 release, the differences have been decreasing.

The IMAGE terrestrial biosphere module is an extended version of the Goudriaan and Ketner model (Goudriaan and Ketner, 1984). The terrestrial biosphere is horizontally divided into seven ecosystems (six in the original model of Goudriaan and Ketner). Vertically, the components biomass (subdivided into leaves, branches, stemwood and roots), litter, humus and charcoal have been distinguished. Furthermore, the deforestation process and its underlying causes have been modelled separately. This is described further in the deforestation module section.

The driving force of the carbon cycle model is the net primary production (NPP), being directed by the fertilization effect. This is the increase in accumulation of carbon in terrestrial ecosystems under influence of higher

3.6. TERRESTRIAL BIOSPHERE MODULE

CO_2 concentrations. In the carbon cycle model this fertilization effect is parameterized by means of a biotic growth factor, even though this factor is ill-known (Goudriaan and Ketner, 1984, Goudriaan, 1987).

The relationship between net primary production and atmospheric CO_2 is generally presumed to be logarithmic (Emanuel et al., 1985, Goudriaan and Ketner, 1984):

$$\frac{dNPP(t)}{NPP(0)} = BIOSTIM * \frac{dpCO_2(t)}{pCO_2(t)} \quad (3.15)$$

with:
$dNPP(t)$ = change in net primary production at time t (GtC/yr)
$NPP(0)$ = initial net primary production, at time $t = 0$, in 1900 (GtC/yr)
$BIOSTIM$ = biotic growth factor; varies in the literature from 0 to 0.7; is 0.4 in IMAGE, whereas Goudriaan and Ketner (1984) take a value of 0.5
$pCO_2(t)$ = atmospheric CO_2 concentration at time t (ppm)
$dpCO_2(t)$ = change in atmospheric CO_2 concentration at time t (ppm)

Recently a new relationship has been incorporated, in which, next to the fertilization effect due to the atmospheric CO_2 increase, the effect of an increase in temperature has also been taken into account, based on Kohlmaier et al. (1990). Both are considered as negative feedbacks.

The carbon flow throughout the system is given in Figure 3.3. The natural input to the system is the net primary production (NPP), while the carbon released upon the decay of the biomass, litter, humus and charcoal components accounts for the natural output of this system. Man-made processes are on the one hand input sources, by humification and charcoal forming, and on the other they are output flows, by changing land use. Both the natural and anthropogenic input and output processes are modelled according to Goudriaan and Ketner (1984):

$$B_{jk}(t) = B_{jk}(0) + \int_{t-1}^{t} [fr_{jk} * NPP(\tau) \\ - \frac{B_{jk}(\tau)}{LF(B_{jk})} - sB_{jk}(\tau) * \sum_{i=1}^{7} a_{ij}(\tau) d\tau] \quad (3.16)$$

with:
$B_{jk}(t)$ = carbon in ecosystem j in component k at time t (in GtC)

$B_{jk}(0)$ = initial carbon in ecosystem j in component k at time $t = 0$, 1900 (in GtC)
fr_{jk} = fraction of NPP partitioned to component k in ecosystem j
$LF(B_{jk})$ = life-span of carbon in ecosystem j of component k (in yr)
$sB_{jk}(t)$ = surface density of carbon in ecosystem j in component k at time t (in GtC/Mha)
$a_{ij}(t)$ = transfer of area from ecosystem j to ecosystem i at time t (in Mha/yr)

$$L_j(t) = L_j(0) + \int_{t-1}^{t} \left[\sum_{k=1}^{3} \frac{B_{jk}(\tau)}{LF(B_{jk})} - \frac{L_j(\tau)}{LF(L_j)} - sL_j(\tau) * \sum_{i=1}^{7} a_{ij}(\tau) d\tau \right] \quad (3.17)$$

with:
$L_j(t)$ = carbon in litter in ecosystem j at time t (in GtC)
$L_j(0)$ = initial carbon in litter in ecosystem j at time $t = 0$, in 1900 (in GtC)
$LF(L_j)$ = life-span of litter in ecosystem j (in yr)
$sL_j(t)$ = surface density of litter in ecosystem j at time t (in GtC/Mha)

3.6. TERRESTRIAL BIOSPHERE MODULE

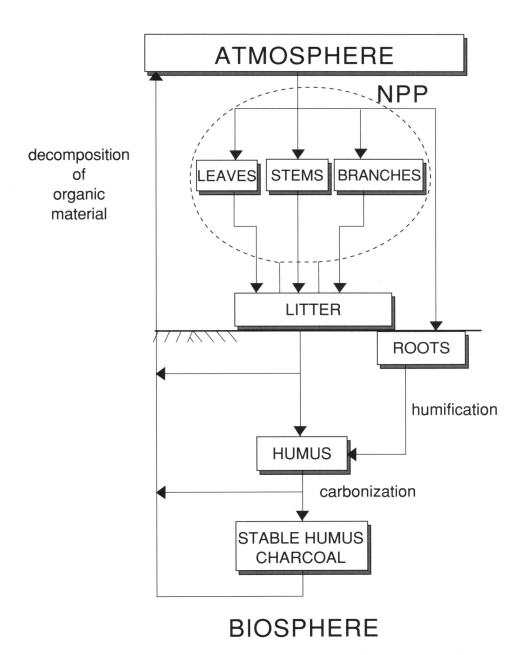

Figure 3.3: Structure of the terrestrial biota model (Goudriaan and Ketner, 1984).

$$H_j(t) = H_j(0) + \int_{t-1}^{t} \left[HF_j * (\frac{L_j(\tau)}{LF(L_j)} \right.$$
$$+ \frac{B_{j4}(\tau)}{LF(B_{j4})}) - \frac{H_j(\tau)}{LF(H_j)} - sH_j(\tau) * \sum_{i=1}^{7} a_{ij}(\tau)$$
$$\left. + \sum_{i=1}^{7} a_{ji}(\tau) * (sH_i(\tau) + HF_j * sB_{i4}(\tau) + 0.5 * sB_{i3}(\tau)) d\tau \right]$$
(3.18)

with:
$H_j(t)$ = humus in ecosystem j at time t (in GtC)
HF_j = humification fraction in ecosystem j
$B_{j4}(t)$ = carbon in roots in ecosystem j (in GtC)
$B_{i3}(t)$ = carbon in stems in ecosystem i (in GtC)
$sH_j(t)$ = surface density of humus in ecosystem j (in GtC/Mha)
$LF(H_j)$ = life-span of humus in ecosystem j (in yr)

$$K_j(t) = K_j(0) + \int_{t-1}^{t} \left[\frac{CFH_j * H_j(\tau)}{LF(H_j)} - \frac{K_j(\tau)}{LF(K_j)} \right.$$
$$- sK_j(\tau) * \sum_{i=1}^{7} a_{ij}(\tau) + \sum_{i-1}^{7} (a_{ji}(\tau) * (sK_i(\tau)$$
$$\left. + \sum_{k=1}^{3} (CFB_k * sB_{jk}(\tau) + CFL * sL_i(\tau)) d\tau \right]$$
(3.19)

with:
$K_j(t)$ = carbon in charcoal in ecosystem j at time t (in GtC)
$K_j(0)$ = initial carbon in charcoal in ecosystem j, at time $t = 1900$ (in GtC)
$LF(K_j)$ = life-span of charcoal in ecosystem j (in yr)
$sK_j(t)$ = surface density of charcoal in ecosystem j at time t (in GtC/Mha)
CFH_j = carbonization fraction of humus in ecosystem j upon decomposition
CFB_k = carbonization fraction of component k upon burning
CFL = carbonization fraction of litter upon burning

3.7. DEFORESTATION MODULE

The land transfer matrix (a_{ij}) links the terrestrial biota module with the deforestation module. While in Goudriaan and Ketner (1984) the matrix elements are related to the growth of world population, in IMAGE these matrix elements are for the greater part dynamic processes, as will be described in the next section, which deals with the deforestation module.

Finally, with the calculated amounts of carbon in the pools of biomass (B_{jk}), humus (H_j), litter (L_j) and charcoal (K_j), the fluxes TNEP and THDIST from Equation 3.2 can be simulated according to Goudriaan and Ketner (1984):

$$NEP_j(t) = NPP_j(t) - (\frac{L_j(t)}{LF(L_j)} + \frac{B_{j4}(t)}{LF(B_{j4})}) * (1 - HF_j)$$
$$- \frac{H_j(t) * (1 - CFH_j)}{LF(H_j)} - \frac{K_j(t)}{LF(K_j(t))} \quad (3.20)$$

with:
$NEP_j(t)$ = net ecosystem production of ecosystem j at time t (in GtC/yr)
$TNEP(t)$ = NEP, summed over all ecosystems at time t (in GtC/yr)
$NPP_j(t)$ = net primary production of ecosystem j at time t (in GtC/yr)

$$HDIST_j(t) = \sum_{j=1}^{7}\sum_{i=1}^{7} a_{ij} * [\sum_{k=1}^{3}(1 - CFB_k) * sB_{jk}(t) - 0.5 * sB_{i3}(t)$$
$$+ (1 - HF_j) * sB_{i4}(t) + (1 - CFL) * sL_i(t)] \quad (3.21)$$

with:
$HDIST_j(t)$ = carbon flux released from biosphere by human disturbance at time t (in GtC/yr)
$THDIST(t)$ = HDIST, summed over all ecosystems at time t (in GtC/yr)

3.7 Deforestation Module

3.7.1 Introduction

The world's tropical forests presently face tremendous pressures from the increasing demands of increasing populations. More than 10 million hectares of closed tropical rainforests are destroyed annually and an equal amount is severely altered. The result of this is the extinction of species, increased erosion, threats to indigenous people, the modification of regional and even global climate and the destruction of a wide variety of possible economically important assets.

Land use change is one of the human activities influencing global change. It can be manipulated in the model by changing the elements of a land use transfer matrix, describing shifts between ecosystems. As a first step towards a global land use change module a tropical forest module was developed to be linked to this matrix in IMAGE. Because the climate effect is only one among several other (possibly more important) effects of deforestation, the module can also be used as a demonstration model on its own. To take other land use changes into account, such as desertification, forest dieback, wetland drainage and extended lowering of groundwater tables, the module has to be upgraded in the future.

Because of the different characteristics of the areas in the module the three major tropical forest areas are considered separately: Latin America, Africa and South-East Asia. It is not intended to present reliable projections of forest resources for the near future, but merely to present long term possibilities for different types of development. In order to calculate long-term deforestation projections, four scenarios were developed: A. unrestricted trends, B. reduced trends, C. changed trends and D. forced trends. From A to D increasing national and international awareness causes the implementation of increasingly protective measures.

In scenario A economic growth is not restricted by environmental concerns. Short time profits prevail over long term assets. For tropical forests this means: a continuing exploitation as if tropical forest were 'mined' as quickly as possible. This means high estimates for logging, agricultural consumption and industrial projects. Because in such a scenario short term economic growth can be considered to generate finances enabling the use of agricultural inputs to increase productivity, the scenario can be expected not to show an exponential trend.

The scenarios B and C are meant to describe two levels of developments in which forest destruction is limited by increasing use of long tern economic potential, as for instance planned in some recent initiatives as the International Tropical Timber Agreement (ITTA) and FAO's Tropical Action Plan. Scenario B is assumed to reflect the effect of the ITTA, whereas in scenario C the Tropical Action Plan of the FAO can be considered to implemented (Food and Agricultural Organization, 1986).

Finally, scenario D is intended to describe the preservation of the forests

3.7. DEFORESTATION MODULE 47

for ecological reasons mainly. Scenario D assumes pure conservation of the forest in their primary state, even if this means, that the economic value of the area would not be used to its full extent. In this scenario, for instance, production of hardwood and meat dairy is limited in the tropical zones and reforestation is important. To simulate scenarios the parameters in the module are varied: demand for wood for construction, fuel wood, energy and minerals, agricultural land and pasture, etc.

Population growth is taken common for all scenarios. Economic development is not included explicitly, since it is assumed that there is no direct easy linkage between economic growth and land use change, and that long term land use change will be driven by the demand for land and products whatever the costs. The details of the different scenarios are described in the Appendix to this chapter.

3.7.2 Description of the Deforestation Model

The deforestation module is an independent submodule within the carbon cycle module. The structure of the deforestation module is shown in Figure 3.4. The module distinguishes four types of ecosystems that are important for the areas concerned: tropical forests, grasslands, arable land and semi-desert. In this study the tropical forest system was split into closed and open forest as defined by Lanly (1982). The ecosystems have different carbon densities for their different components (vegetation, litter, soil humus).

Intermediate types of forests (secondary forests, forest fallow) were included in the 'open forest' system. All converted forests are assumed to end up as re-established forests, agricultural or pasture land, or degraded area. Deforestation is triggered by a variety of processes, which are mainly caused by a number of demands driven by growth of population and economy.

Notation:

The ecosystems (in ha) and processes (in ha/yr) are denoted as follows:

CLSF	=	closed tropical forest				
OPNF	=	open tropical forest				
AGRIC	=	arable land				
GRASS	=	pasture land				
OTHR	=	tundra and semi-desert land				
DEFAGC	=	expansion of arable land into closed forest	from	CLSF	to	AGRIC
DEFAGO	=	expansion of arable land into open forest	from	OPNF	to	AGRIC
DFLIVC	=	expansion of pasture into closed forest	from	CLSF	to	GRASS
DFLIVO	=	expansion of pasture into open forest	from	OPNF	to	GRASS
YSHFTC	=	shifting cultivation in closed forest	from	CLSF	to	OPNF
DEFYSH	=	shifting cultivation in open forest	from	OPNF	to	AGRIC
YDMOTH	=	expansion of agricult. land into other land	from	OTHR	to	AGRIC
COMM	=	logging by commercial wood production	from	CSLF	to	OPNF
REFOR	=	development of plantations	from	OTHR	to	OPNF
FWOOD	=	fuelwood gathering	from	OPNF	to	OTHR
IND	=	industrial or mining projects	from	CLSF	to	OTHR
DEGAGR	=	degradation of arable land	from	AGRIC	to	GRASS
DEGGRA	=	degradation of pasture	from	GRASS	to	OTHR
GRDSRT	=	desertification of grassland	from	GRASS	to	OTHR
AGDSRT	=	desertification of agricultural land	from	AGRIC	to	OTHR

In the deforestation computer model several processes (DFLIVC, DFLIVO, DEFAGC, DEFAGO, DEFYSH and YDMOTH) may become negative. Therefore these processes are defined in such a way that, in case of the occurrence of a negative value, this process is set to 0 and the reverse process is activated. For instance DFLIVO (deforestation of open forest by livestock, from OPNF to GRASS) has an analogously defined reverse process DFLIOR (reversed deforestation, from GRASS to OPNF), which is 0 if DFLIVC is greater than 0 and receives the absolute value of DFLIVC if DFLIVC is less than 0. The same holds for the other processes DFLIVO, DEFAGC, DEFAGO, DEFYSH and YDMOTH.

The land use changes resulting from the calculations are represented by the already mentioned processes and are inserted at the relevant places of the transfer matrix as shown in Table 3.1.

3.7. DEFORESTATION MODULE

from \ into	closed forest	open forest	temperate forest	grass-land	agric. land	human area	semi-desert tundra
closed forest	10	0	0	0	0	0	0
open forest	YSHFTC + COMM	6.5	0	0	DEFYSH	0	REFOR
temperate forest	0	0	2	0	0	0	0
grassland	DFLIVC	DFLIVO	1	400	DEGAGR	0	0
agricult. area	DEFAGC	DEFAGO + DEFYSH	0	0	400	0	YDMOTH
human area	0.5	0	0.5	1	1	0	0
semi-desert/tundra	IND	FWOOD	0	DEGGRA + GRDSRT	AGDSRT	0	0

Table 3.1: Modified process transfer matrix of area between ecosystems (Mha/yr) in 1980.

As can be seen in Figure 3.4 some simplifications have been made. Fuelwood gathering is only taken into account for open forests, assuming that in closed forest areas the natural production capacity surpasses local demand (de Montalembert and Clément, 1983). Forest plantations are assumed to develop on other lands rather than on agricultural, grass or forest lands. Logging is assumed to take place in closed forests only, degrading them into secondary (here open) forests. These processes will be discussed in the following sections. To validate the different relationships for the past decades the FAO Production Yearbooks were used as well as a number of literature estimates of present and historical values of important parameters, among which are those cited by the World Resources Reports (World Resources Institute, 1986, 1987, 1988/1989). The most important assumptions and input data are summarized in the Appendix to this chapter.

Land transfers are described by differential equations, while processes are represented by algebraic dynamic functions. The amount of closed tropical forest is determined by the initial amount and the conversion into arable land (process DEFAGC), pasture (process DFLIVC), other land (process IND) and open tropical forest (processes COMM and YSHFTC), respectively. This leads to the following dynamic expression:

$$CLSF(t) = CLSF(t-1) + \int_{t-1}^{t} (-COMM(\tau) - IND(\tau)$$
$$- YSHFTC(\tau) - DEFAGC(\tau) - DFLIVC(\tau))d\tau$$
(3.22)

with:
$CLSF(0) =$ amount of closed forest at time 0, in 1900 (in ha)

Similar to closed forest, the amount of open tropical forest is determined by the initial amount and the conversion into arable land (processes DEFAGO and DEFYSH), pasture (process DFLIVO), other land (FWOOD) and the re-establishment of open forest (process REFOR); additionally closed forests are converted into open forests (processes COMM and YSHFTC). Hence the dynamic amount of open tropical forest is expressed as:

$$OPNF(t) = OPNF(t-1) + \int_{t-1}^{t} (COMM(\tau) + REFOR(\tau)$$
$$+ YSHFTC(\tau) - DEFAGO(\tau) - DFLIVO(\tau)$$
$$- FWOOD(\tau) - DEFYSH(\tau))d\tau \qquad (3.23)$$

with:
$OPNF(0) =$ amount of open forest at time 0, in 1900 (in ha)

The penultimate ecosystem to be described is arable land. The amount of arable land is specified by the initial amount and the expansion of agricultural land into closed forest (processes DEFAGC) and open forest (processes DEFAGO and DEFYSH) and into other land (process YDMOTH) minus the degradation of arable land to grassland (process DEGAGR):

$$AGRIC(t) = AGRIC(t-1) + \int_{t-1}^{t} (YDMOTH(\tau) - DEGAGR(\tau)$$
$$+ DEFAGC(\tau) + DEFAGO(\tau) + DEFYSH(\tau))d\tau$$
(3.24)

with:
$AGRIC(0) =$ amount of agricultural land at time 0, in 1900 (in ha)

3.7. DEFORESTATION MODULE

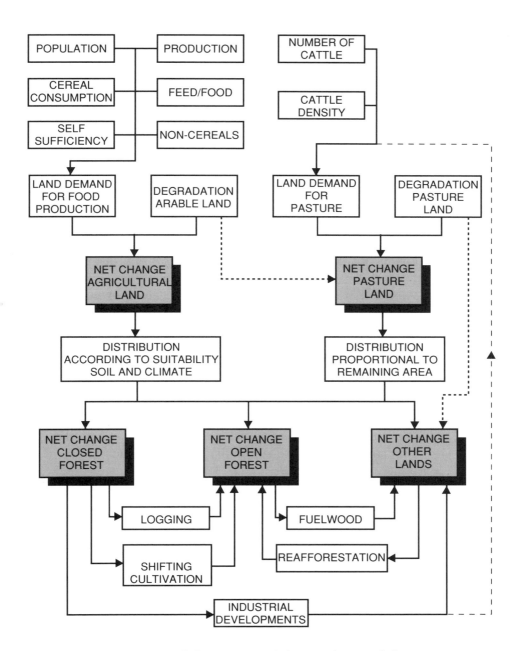

Figure 3.4: Structure of the IMAGE deforestation module

Finally the ecosystem pasture (grassland) is defined, specified by the initial amount of pasture, the conversion of tropical forest into pasture (processes DFLIVC and DFLIVO) and degradation of arable land to grassland (DEGAGR), minus the degradation of grassland to other land (DEGGRA):

$$\begin{aligned} GRASS(t) = {}& GRASS(t-1) + \int_{t-1}^{t} (DEGAGR(\tau) - DEGGRA(\tau) \\ & + DFLIVC(\tau) + DFLIVO(\tau)) d\tau \end{aligned} \quad (3.25)$$

with:
$GRASS(0) =$ amount of pasture (grassland) at time 0, in 1900 (in ha)

A few of the most important processes are elucidated below. These processes are represented by algebraic equations. The other processes are expounded in Swart and Rotmans (1989a,b).

The expansion of agricultural land into closed tropical forest is represented by the demand for arable land and the degradation of arable land, multiplied by a suitability factor, representing the proportion between closed forest, open forest and other land within a suitability class (for an explanation of the term suitability class see §3.7.3)). This agricultural expansion is defined by the following algebraic equation:

$$DEFAGC(t) = (YDMAGR(t) + DEGAGR(t)) * ALFCLF(t) \quad (3.26)$$

with:
$YDMAGR(t) =$ yearly demand for arable land at time t (in ha/yr)
$ALFCLF(t) \;\;=$ ratio of current amount of closed tropical forest to the current amount of closed forest, open forest and other land according to suitability distribution at time t

Exactly the same kind of equation holds for the expansion of open tropical forest into agricultural land (Swart and Rotmans, 1989a,b). Analagous to the agricultural expansion, the expansion of pasture into closed forest is determined by the demand for grassland, the degradation of both grassland and arable land, multiplied by a ratio, reflected in the following expression:

$$\begin{aligned} DFLIVC(t) = {}& (YDMGR(t) + DEGGRA(t) \\ & - DEGGRA(t)) * ALFGRC(t) \end{aligned} \quad (3.27)$$

with:
$YDMGR(t) =$ yearly demand for pasture at time t (in ha/yr)
$ALFGRC(t) =$ ratio of current amount of closed tropical forest to the initial amount of closed tropical forest.

3.7. DEFORESTATION MODULE

The same applies to the expansion of pasture into open forest. The expansion of agricultural land into other land is defined as a function of the yearly demand for agricultural land multiplied by a ratio:

$$YDMOTH(t) = YDMAGR(t) * ALFOTH \qquad (3.28)$$

with:
$YDMOTH(t)$ = colonization (yearly expansion of agricultural land into other land) at time t (in ha/yr)
$YDMAGR(t)$ = yearly demand for agricultural land at time t (in ha/yr)
$ALFOTH(t)$ = ratio of the current amount of other land to the current amount of closed forest, open forest and other land in the current suitability class at time t

For an exhaustive and more detailed description of the other driving forces and processes of the deforestation module is referred to Swart and Rotmans (1989b).

3.7.3 Description of the Processes

The following processes will be described: permanent agriculture, shifting or pioneer cultivation, cattle breeding, logging, fuelwood gathering, industrial projects, and reforestation.

Permanent agriculture

Mankind has always transformed the face of the earth since prehistoric times. The scope of this transformation has accelerated since man's lifestyle changed from nomadic to sedentary, culminating in the spectacular population growth after the industrial revolution (Wolman and Fournier, 1987). Expansion of permanent agriculture has always been the most important factor in deforestation at all latitudes. Presently the demand for agricultural land in the tropical regions is primarily caused by five factors:

- an increasing number of people

- higher levels of consumption of food, fiber and forest products due to economic growth

- area and productivity loss by land degradation

- inaccessibility of intensive agriculture for poor farmers
- production for export (often for debt relief).

Many studies have indicated that, theoretically, present and future world populations can be amply fed by increases in arable land and productivity (e.g. Buringh et al., 1975). Other studies have indicated that this is not so for individual countries (FAO, 1984). Unfortunately, in past studies on population carrying capacities the present vegetation has not been taken into account. In this study it is. To determine the net demand for additional agricultural land for each region the following parameters have been taken into account: population, cereal demand per capita, cereal productivity per hectare, present arable land, fraction of land used for non-cereals and other products, fraction of cerals used for feed, the self-sufficiency ratio to allow for food imports or aid.

$$DMAGR(t) = \frac{POP(t) * CE(t) * SSR(t)}{Y(t) * \alpha(t) * (1 - \beta(t) - \gamma(t))} \tag{3.29}$$

with:
$DMAGR(t)$ = cumulative demand for agricultural land at time t (in ha)
$POP(t)$ = population at time t
$CE(t)$ = cereal consumption at time t (in kg/cap)
$SSR(t)$ = self sufficiency ratio at time t
$Y(t)$ = yield at time t (in kg/ha)
$\alpha(t)$ = fraction of cereals used for food at time t
$\beta(t)$ = fraction of area for food from non-cereals at time t
$\gamma(t)$ = fraction of area for other other agricultural products at time t

To determine the total demand for agricultural land finally degraded lands are added. For lack of quantitative information degradation is simulated by a negative exponential function, which encompasses the assumption that every newly converted hectare will be more susceptible to erosion than the previous one. High inputs are not only assumed to increase agricultural productivity, but also to decrease erosion rates. Additional effort is necessary for a better quantitative estimate of land degradation.

The demand for agricultural land has to be distributed over the different ecosystems. The following procedure has been followed. First, those areas are converted which have been altered by logging or shifting cultivation the year before. Then the land is assumed to be colonized according to its

3.7. DEFORESTATION MODULE

suitability for agriculture. For this project, estimates of the suitability of areas have been made by Bouwman (1989a,b), taking into account the soil types, the climate, the topography and the input level, see Table 3.2. For scenarios A and B high input levels have been assumed, and low levels for C and D.

mln ha	suitability classes[1]								
	low input				high input				
Africa	I	II	III	IV	I	II	III	IV	total
closed forest	.7	27.8	151.0	41.7	43.3	149.3	5.8	22.7	221.1
open forest[2]	1.0	136.9	165.8	194.8	162.5	227.5	30.0	78.6	498.5
grassland	6.6	186.0	170.0	415.7	221.8	233.0	35.1	288.3	778.2
arable land	1.1	50.5	64.4	67.1	67.7	76.1	12.4	27.0	183.1
semi-desert[3]	2.8	82.7	93.7	1098.0	111.7	119.4	21.9	1024.2	1277.1
total	12.1	483.8	644.7	1817.3	606.9	805.3	105.1	1440.7	2958.0
Latin America									
closed forest	0.0	53.1	509.4	130.7	403.3	179.8	66.0	44.0	693.2
open forest[2]	0.0	46.3	87.3	106.0	86.91	106.2	9.6	36,8	239.5
temperate forest	0.0	22.0	6.6	14.7	22.7	0.0	11.1	2.9	36.7
grassland	5.3	97.8	202.1	245.2	173.1	201.8	57.2	118.3	550.4
arable land	0.0	51.2	85.8	38.5	108.3	38.3	16.7	12.2	175.5
semi-desert[3]	0.6	57.0	18.3	215.8	58.4	30.5	29.3	173.3	291.6
total	5.9	327.3	902.8	750.8	852.7	556.7	190.0	387.5	1986.9
Southeast Asia									
closed forest	7.8	62.2	179.6	55.9	220.6	29.6	35.9	19.4	305.5
open forest[2]	0.0	15.6	3.5	11.9	25.0	1.5	3.4	1.1	31.0
grassland	0.0	4.3	9.9	16.0	12.5	3.5	5.0	9.2	30.2
arabel land	0.8	97.5	55.0	103.9	166.6	20.5	25.8	44.3	257.2
semi-desert[3]	1.7	53.8	58.2	71.5	119.8	15.1	22.0	28.3	185.0
total	10.2	233.3	306.2	259.1	544.4	70.2	92.1	102.2	808.8

1 I=high, II=moderate, III=low and IV=marginal suitability

2 including forest fallow, logged forest, forest plantations

3 including human area. Shifting or pioneer cultivation

Table 3.2: Agricultural suitability classes (derived from Bouwman, 1989a,b).

Shifting or pioneer cultivation

Much controversy exists on the subject of shifting cultivation. According to Lanly (1982) shifting cultivation is increasing because of increasing numbers of people; according to Myers (1984) it is decreasing, because landless farmers from other areas convert traditional shifting cultivation areas into permanent cropland. For the carbon cycle a sustainable traditional form of shifting cultivation is not very important. Without entering the discussion on the definition of 'shifting' or 'pioneer' cultivation the following approach is chosen: expanding traditional shifting cultivation in closed forest areas is considered as a catalysing agent for further forest destruction in later years, degrading them initially into open forests. First the expansion into closed forests is calculated as a function of the fraction of the rural population involved in this type of agriculture, the fallow period and the conversion per family (Detwiler et al., 1985).

$$\begin{aligned}SHFTCL(t) = & POP(t) * RUR(t) * SWFC(t) \\ & *(CLF(t)/NFM(t)) * FC(t)\end{aligned} \quad (3.30)$$

with:
$SHFTCL(t)$ = area under shifting cultivation at time t (in ha)
$POP(t)$ = population number at time t
$RUR(t)$ = rural fraction at time t
$SWFC(t)$ = rural fraction shifting cultivation (swidden farmers) at time t
$CLF(t)$ = amount of land cleared per family per year at time t (in $ha/fm.yr$)
$NFM(t)$ = number of people per family at time t (in cap/fm)
$FC(t)$ = fallow cycle at time t (in yr)

In the computer program this converted area is memorized and in the next timestep it is first used for expansion of permanent agricultural area.

Cattle breeding

In all developing countries the number of cattle has gradually increased along with the population. Only in Latin America has this growth taken place at the expense of tropical forests. In Asia and Africa meat and dairy products form a less important part of the diet, while no major export markets are near at hand. The troublesome situation in Latin America, where cattle breeding is expanding, originally triggered by beef demands in the USA and maintained by a complex of socio-economic factors, is discussed extensively in the literature (Hecht, 1988, Fearnside, 1987 or Repetto, 1988). In Asia, cattle are integrated in the agricultural system and in general do not require separate grazing lands. They feed on agricultural wastes and off-season feed products.

At this point in time it is not clear whether this situation might change in the future. When economies surpass a certain welfare level, the demand for luxury products like meat and dairy may jump to a much higher level, which might not be served by the present integrated situation, but might

3.7. DEFORESTATION MODULE

require additional land for grazing and feed production. In Africa a different situation exists. Many areas are less suitable for agriculture than for cattle breeding. Therefore livestock is an important source of income for many people. Although the extensive cattle breeding has expanded in Africa, it appears that the most important problem is the land degradation by overgrazing rather than forest destruction. More study on livestock in these two regions is needed. For the different scenarios different growth rates of numbers were taken and different levels of intensity (numbers per hectare):

$$DMGRS = GRSSIN * CAT/CT1900 * CATPRD\ PR1900 \qquad (3.31)$$

with:
$DMGRS(t)$ = cumulative demand for grassland (in ha)
$GRSSIN(t)$ = initial amount of grassland (in ha)
$CAT(t)$ = number of cattle
$CT1900(t)$ = number of cattle at time $t = 0$, in 1900
$CATPRD(t)$ = cattle productivity (in ha/hd)
$PR1900(t)$ = cattle productivity at time $t = 0$, in 1900 (in ha/hd)

Logging

In public discussions on tropical deforestation in the developed world production of commercial tropical hardwood is often pointed at as the most important cause of deforestation. The wood traders then reply that they do not cause more than a few percent of the total. Both may be right. Directly, commercial logging may not contribute much in comparison with other causes. But indirectly, logging activities open up the forests by road construction, taking poor farmers in their wake to finish the job. In this model this effect is simulated in a similar way to shifting cultivation: after logging in the next timestep the logged area is converted into agricultural area. This assumption is not valid everywhere, for instance in parts of Indonesia, where logs are transported by river. Generally logging can be considered as the mining of a non-renewable resource. Commercially preferred species need 40 to 100 years for full regrowth, were they to be replanted, which does not happen yet.

The per capita demand for wood and wood products is more or less proportional to economic growth. This means that, in the future, an increasing portion of the wood production of the tropical forest areas will be used internally. Even now some tropical forest countries are wood importers. Supplies in the developed world will be restricted by the applied principle of sustained yield and maybe by the effects of acidification. Now that the forest resources of Southeast Asia and West Africa are becoming depleted, attention will be focused on hitherto relatively unexploited regions like Amazonia and the Congo basin. There are a number of ways to reduce the stress on wood resources. First, more species have to be utilized. This does not stop the 'mining' aspect of tropical wood production, however. More efficient end use or the use of alternative materials is needed. Then the development of tropical hardwood plantations has to be stimulated. In the model scenarios two parameters are varied: the growth rate of hardwood production and the productivity per hectare. The simulated logging for the three regions is plotted in Figure 3.5.

Fuelwood gathering

Combustion of fuelwood only contributes to the greenhouse effect when extraction is larger than regrowth. This is the case in many regions, where fuelwood scarcities occur (de Montalembert and Clément, 1983). Usually these areas lie outside the major forest areas. Lanly (1982) considers fuelwood gathering in forests as a less important degradation factor rather than a cause of total forest destruction. Fuelwood is the primary source of energy in rural areas. Since scarcities arise in the rural areas and wood is transported over ever longer distances, commercial fuelwood extraction from forests may increase, leaving large areas of forest severely damaged. According to Myers (1984) the demand for fuelwood already causes major deforestation (25,000 km^2 annually). For this study it is assumed that closed forests are not affected because of natural regrowth. Only open forests are taken to be deforested, at a rate which is chosen proportional to rural population growth in the area. No allowance for decreasing withdrawal because of plantation wood or increased combustion efficiency has yet been taken into account.

3.7. DEFORESTATION MODULE

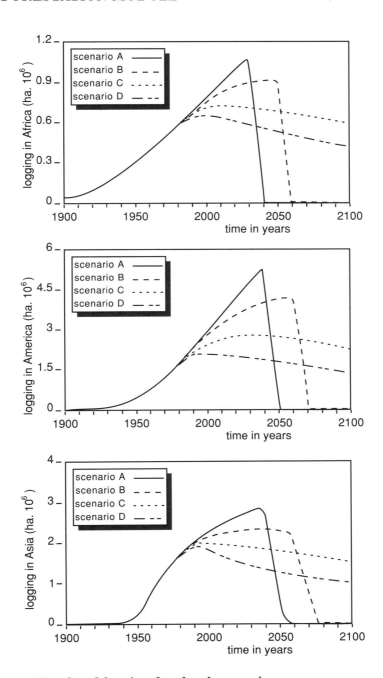

Figure 3.5: Simulated logging for the three regions.

Industrial projects

Industrial projects are seldom considered in systems studies on deforestation. The implementation of projects is primarily steered by political and financial aspects. Although the direct effect is usually rather small in terms of hectares of forests cleared, the opening up of virgin land and possible employment opportunities draw vast amounts of people to the project area. Although in all three regions these projects play a role, only the massive plans for Amazonia have been included, aiming at the exploitation of the mineral resources and hydropower potential. The planned size of the projects depends very much on government changes and opportunities for external financing, and changes rapidly. In order to illustrate the potential importance of these projects, different rates of forest conversion are included for the different scenarios for Latin America.

Forest plantations

Next to the abatement of deforestation, re-afforestation is often brought forward as a possible instrument to combat climate change. The calculated areas necessary to balance fossil fuel emissions depend very much on the assessment of the productivity of forest plantations. Optimists use figures taken from relatively small, well-managed plots, while pessimists argue that the associated productivities cannot be applied to the large-scale plantations necessary to have a significant impact on the carbon cycle.

Furthermore, in the developing countries forest plantations will be established for more pressing needs than carbon sequestering, such as fuelwood supply or erosion control. Re-afforestation of abandoned agricultural lands or other lands in the developed countries may have significant potential but is not covered by this module. In the most optimistic scenario present plantation rates as given by Lanly (1982) are used and the areas increase logistically to 345 million hectares in Africa, 310 million hectares in Latin America and 100 million hectares in Southeast Asia. In the other scenarios only 75, 50 and 25%, respectively, of these areas would be afforested, see Figure 3.6.. Since this variable is not integrated with the energy module, the positive contribution to the greenhouse effect stops when the final forest areas are reached. Theoretically, forest plantations could have a longer-lasting effect when short rotation plantations are used to provide fuelwood as a replacement for fossil fuels.

3.7. DEFORESTATION MODULE

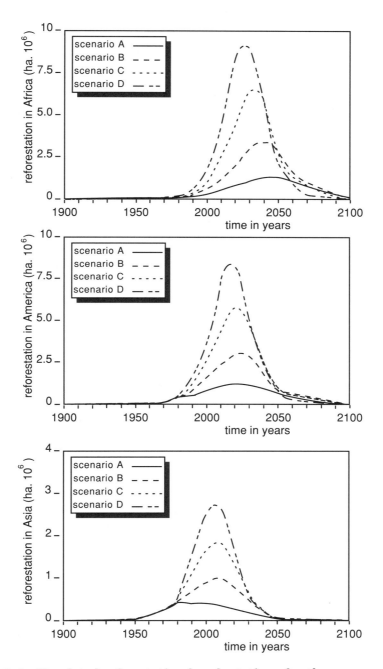

Figure 3.6: Simulated reforestation by plantations for three areas.

3.8 Validation and Uncertainty

There are two ways of validating the model results. First, to compare the simulated CO_2 concentrations with measured concentrations at Mauna-Loa, for the known series of measurements from the year 1960 to 1985 (Keeling et al., 1982). A comparison of the simulated concentrations with measured concentrations yields a maximum relative difference of 1.4% in the year 1960 and a minimum relative difference of 0.05% in 1985. Analyzing the simulation pattern of historical CO_2 concentration shows a slightly more rapid increase in the calculated CO_2 concentration compared with the increase in the Mauna-Loa concentration series.

Secondly, for validation of the different ecosystem areas and rates of conversion the statistical information of the FAO was used, together with a limited number of estimates from the literature. With the present deforestation system dynamics it was difficult to simulate values for conversion of tropical forests below 20%. Therefore estimates below 10% seem to be improbable, especially for Southeast Asia. More historical evidence of land use changes will improve these insights. Simulation results with the deforestation module of IMAGE for the period 1900 to 1985 show a decline in the tropical forest areas of 20 to 25%, which is higher than some estimates in the literature.

Uncertainties in this carbon cycle model arise from the choice of vegetation types, being rather arbitrary, the biotic feedback on atmospheric CO_2 increase (fertilization effect), represented by the biotic growth factor, and the role of charcoal production by burning. The role of the latter in the carbon cycle is especially controversial (Jansen, 1987).

In view of these uncertainties it is of crucial importance to perform sensitivity and uncertainty analyses with these carbon cycle models. Examples of such analyses are presented in USDOE (1985a,b) and Yearsley and Lettenmaier (1987). A description of a sensitivity analysis applied to the carbon cycle module is described in Chapter 12.

3.9 Results

The airborne fraction, as defined in (3.1), is represented in Figure 3.7. In scenario D a continuous decrease takes place, down to about 25% in 2100, while for scenario A an increasing value is reached of about 70%. The near equilibrium between the uptake of CO_2 (by oceans and terrestrial biota) and the release of CO_2 (by fossil fuels and land use change) in scenario D,

3.9. RESULTS

causes a low airborne fraction. From the results of the calculations with the carbon cycle module the biosphere appears to have been an important carbon source in the past, but a minor sink at present. Although still resulting in a positive carbon flux from the biosphere, estimates of other authors also show a declining trend (Houghton et al., 1983, 1985 and 1987, Detwiler and Hall, 1985, 1988).

Concerning the greenhouse problem the items referred to most are the time at which a doubling of the initial atmospheric CO_2 concentration will occur and the consequent temperature increase. Figure 3.8 shows that doubling of the CO_2 concentration will be reached in about 2060 (scenario A), 2075 (scenario B), 2100 (scenario C) or even after 2100 (scenario D).

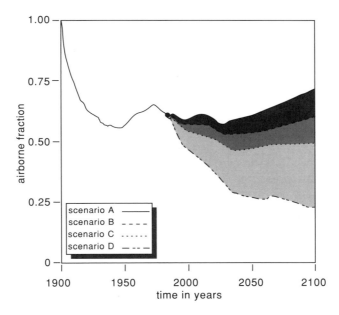

Figure 3.7: Airborne fraction.

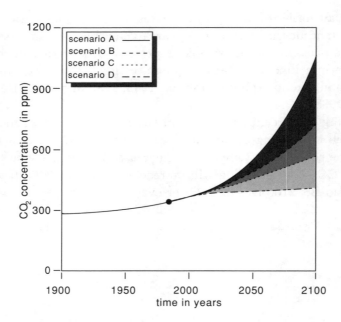

Figure 3.8: Concentration of CO_2.

In Figures 3.9 and 3.10 the remaining amounts of closed and open forests for four scenarios are presented for the three regions. In the most pessimistic scenario the closed tropical forests would disappear before the middle of the next century. Although one may expect the forests in the remote Amazon and Congo basin to last the longest, we find that they may disappear at least as fast as the Asian forests. This result can partly be explained by the fact that we used estimates of past land use changes by Houghton et al. (1983) to validate our results for the period 1900–1980. This source suggests that in Asia only a small percentage of the tropical forests would have been destroyed. Very likely this finding is not correct, taking into account that in several Southeast Asian countries 30 to 40% of the forests have disappeared (World Resources Institute, 1988). In this model this also affects the future deforestation rates. The curves for open forests show upward trends for some scenarios, which is primarily caused by the inclusion of forest plantations in this category.

Figures 3.11 and 3.12 show the causes of deforestation as simulated in 1980 and one particular future year, namely 2020. For closed forests the demand for agricultural land, pasture and commercial wood are of similar importance, while for open forests both agriculture and fuel wood demand

3.9. RESULTS

play a dominant role. The demand for agricultural land and pasture is partly driven by increasing loss of land by erosion. To avoid the large increases in agricultural land in forest area dramatic increases in productivity in other areas is necessary.

In Figure 3.13 the interesting increases for agricultural land are shown as simulated by the model. In the higher scenarios increases in demand driven by growth of economy and population exceed the effect of increasing yields. In scenario A the demand for agriculture and pasture not only requires the conversion of all forests but also a significant fraction of the 'other lands'. Further elaboration of these results will be useful.

The important role of soil degradation is emphasized by Figure 3.14, showing the degradation of arable land. The simulated loss of arable land increases in both absolute and relative terms. This is caused by the fact that in the degradation relation the increase of degradation by expansion of land dominates over the decrease of degradation by input increases.

In Figures 3.15 and 3.16 the impact of different deforestation scenarios on the atmospheric CO_2 concentration is shown. In Figure 3.16 a common forced trend energy scenario is taken for both lines. The difference between deforestation scenario A and D is limited to about 50 *ppm*, or roughly 10%. For the higher energy scenarios the relative differences are slightly smaller, as Figure 3.15 shows for an unrestricted trends energy scenario. For this scenario the difference between deforestation scenario A and D is limited to about 50 *ppm* CO_2, which is about 5%. Earlier, around 2050, the relative difference is larger because the logistic approach for afforestation leads to the higher rates of carbon sequestering in the first half of the next century. These findings indicate that deforestation is important, but small in comparison to the contribution of fossil fuel combustion to the greenhouse effect, see Swart and Rotmans (1990). It is important to note that this result is not the separate and direct effect of the destruction of tropical forests alone, but the net impact of all changes of land use and atmospheric composition. In the dynamics of the carbon cycle by Goudriaan and Ketner (1984) the CO_2 release by deforestation is counteracted by other factors like the CO_2 fertilization effect and charcoal formation after burning of vegetation. An elaborate sensitivity analysis shows that in the C cycle model the factors concerning land use change are of moderate importance in comparison with dominating factors as the fertilization factor or the grassland humification factor (Swart and Rotmans, 1989b). This accounts to a large extent for the limited impact of deforestation in the Goudriaan and Ketner carbon cycle module.

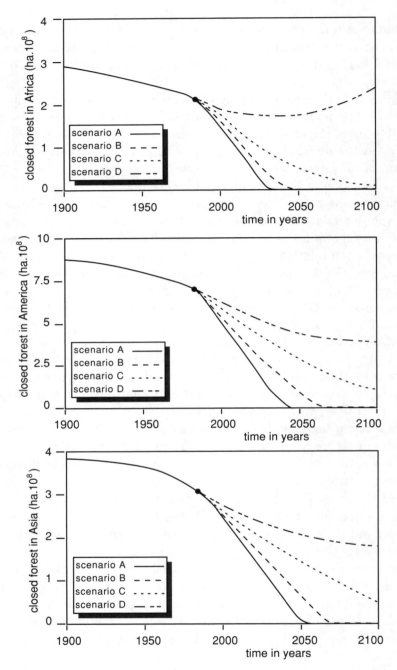

Figure 3.9: Simulated areas of closed tropical forests for three regions.

3.9. RESULTS

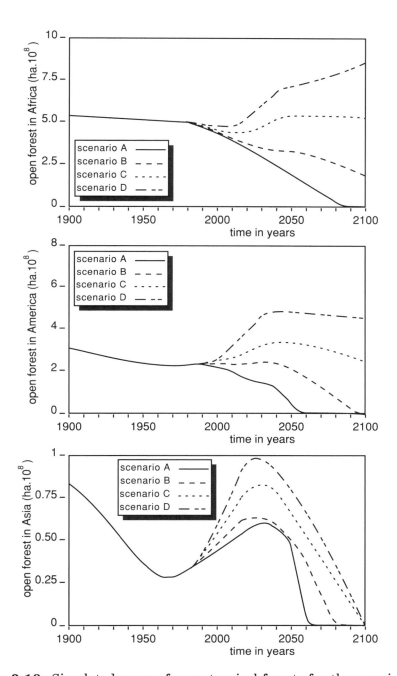

Figure 3.10: Simulated areas of open tropical forests for three regions.

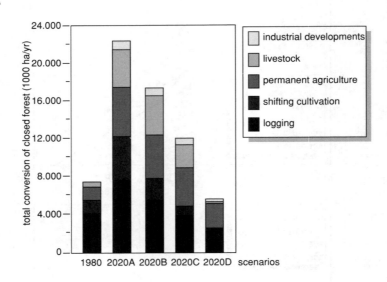

Figure 3.11: Conversion of closed tropical forests by cause for selected years.

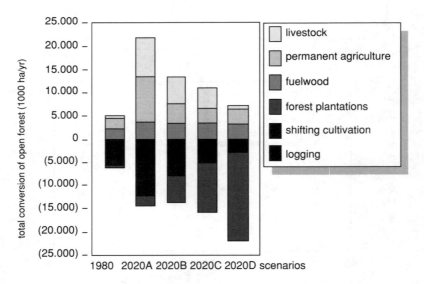

Figure 3.12: Conversion of open tropical forests by cause for selected years.

3.9. RESULTS

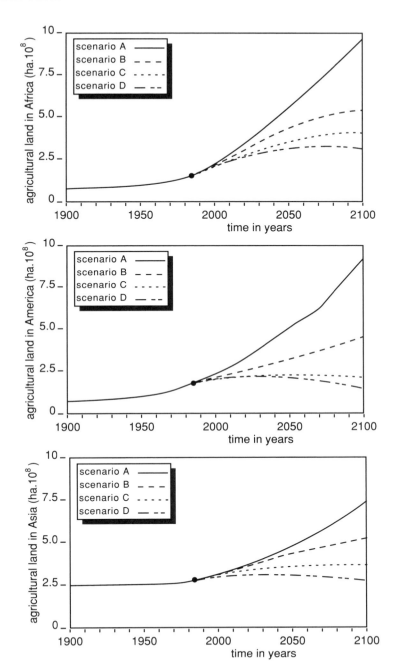

Figure 3.13: Simulated areas of agricultural land for three regions.

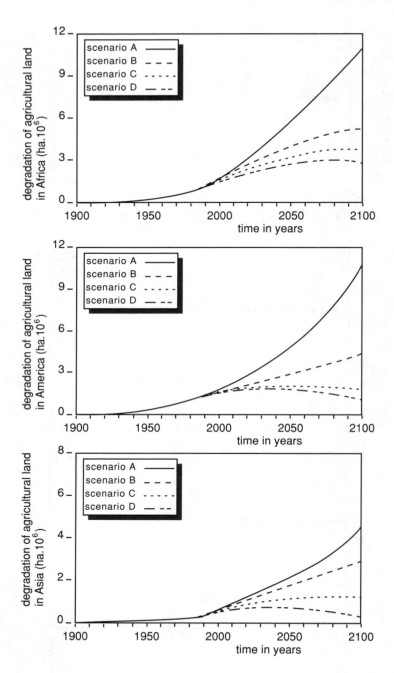

Figure 3.14: Simulated degradation of agricultural land for three areas.

3.9. RESULTS

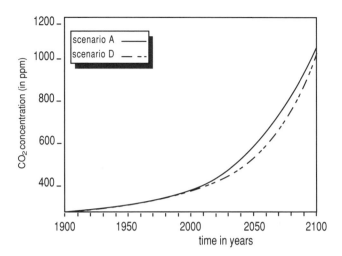

Figure 3.15: Simulated atmospheric CO_2 concentrations for the highest and lowest deforestation scenario combined with an unrestricted trends energy scenario

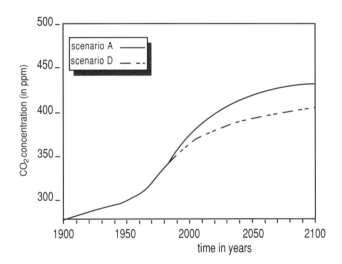

Figure 3.16: Simulated atmospheric CO_2 concentrations for the highest and lowest deforestation scenario combined with forced trend trends energy scenario

3.10 Conclusions

o Continuation of the recent trend of emissions of CO_2 leads to a rapid rise of the atmospheric CO_2 concentration. Even radical measures (scenario D: forced trends scenario) cannot prevent the CO_2 concentration from increasing, although in this case a CO_2 doubling is not reached before 2100.

o The contribution of biospheric changes to the greenhouse effect is important but small as compared to that of fossil fuel combustion. Model simulations with IMAGE show a difference of a maximum of 10% in CO_2 concentrations over the next century between optimistic and pessimistic deforestation cases. If not for reasons of global climate change, then deforestation should be stopped for other reasons, including the combat of erosion, the conservation of species diversity, safety of indigenous people and avoidance of local and regional climate changes.

o The most important direct cause of deforestation is the demand for agricultural land to satisfy demand for food, feed or debt-resolving export products. The combination of growth of population and their consumption level will increase the pressure on the forests if no drastic corrective measures are taken.

o Simulation results with IMAGE show that if present deforestation rates continue a total destruction of the tropical forests will occur halfway through the next century. This would be disastrous although its effect on global climate is only one reason to stop tropical deforestation. Other possibly more important reasons are the inefficient use of the resource leading to erosion, the loss of species, the threat to the indigenous population and the effects on local and regional climate.

o Although it is difficult to give quantified estimates, IMAGE simulations show that soil degradation is a process that, while seldom mentioned in this connection, contributes considerably to the rate of tropical deforestation by decreasing the availability of land for agriculture and pasture.

o The result that even rapid deforestation has only a moderate net effect on the carbon cycle is in IMAGE primarily caused by an increasing soil carbon pool due to the conversion of organic matter into charcoal during burning, by CO_2 fertilization and to possible underestimation

3.10. CONCLUSIONS

of soil losses after conversion. To some extent the biospheric CO_2 emissions by deforestion will be counteracted by a CO_2 fertilization effect. However, the less vegetation that remains, the less potential there is for this effect.

o The IMAGE deforestation module shows that the development of a global land use change model is feasible and can provide valuable insights into the interactions between human activities and the earth system. Such a model, possibly in combination with a long term energy model, could not only be used to study the impact of land use changes on climate, but also on other important biogeochemical cycles.

o Although reforestation can play a significant role in a transition period, a gradual shift towards a world energy system based on renewables away from fossil fuels is a more effective strategy for the mitigation of climatic change.

o Large scale industrial, infrastructural, mining or livestock projects have an enormous potential for accelerating the deforestation process, which is, however, difficult to quantify. Integrated economic and ecological evaluation is crucial to avoid unnecessary ecological disasters.

o Decrease of agricultural area and productivity by land degradation is an important factor, catalysing deforestation. Erosion control and conversion of soils less susceptible to erosion should have priority. The importance of this factor warrants a more thorough analysis of this process than has been applied in this study.

3.11 Appendix

IMAGE DEFORESTATION MODULE DATA SHEET

Africa		in 2100 scenario to	
	unit	1900	1985
remaining closed forest	1000 ha	294000	221079
remaining open forest	1000 ha	539000	498479
fallow tot. AF closed	1000 ha		61646
swidden cycle closed	years	10	10
fallow tot. AF open	1000 ha		104335
fallow period open	years		15
population function[1]	mln		555
percentage rural	%	70	70
swidden farmers closed	%	2.5	4
swidden farmers open	%	10	10
number in family	cap/fm	8	7
cleared per family	ha/yr	1.3	1.3
human area (0.05 ha/cap)	1000 ha		27750
yield tot. cereals sc. A^1	kg/ha		800
yield tot. cereals sc. B^1	kg/ha		800
yield tot. cereals sc. C^1	kg/ha		800
yield tot. cereals sc. D^1	kg/ha		800
consumption food cereals	kg/cap	110	110
percentage food (vs. feed)	%	100	84
area non-cereals	%	61	61
self-sufficiency ratio[2]	%	100	79
forest plantations sc. A^1	1000 ha	1	2411
forest plantations sc. B^1	1000 ha	1	2411
forest plantations sc. C^1	1000 ha	1	2411
forest plantations sc. D^1	1000 ha	1	2411
cattle number scen. A^1	1000	60000	140000
cattle number scen. B^1	1000	60000	140000
cattle number scen. C^1	1000	60000	140000
cattle number scen. D^1	1000	60000	140000
cattle productivity	ha/hd	8.0	4.5
comm. wood prod. index^{1,4} A	ha/yr	50400	639000
comm. wood prod. index^{1,4} B	ha/yr	50400	639000
comm. wood prod. index^{1,4} C	ha/yr	50400	639000
comm. wood prod. index^{1,4} D	ha/yr	50400	639000
wood prod. int. index		0.09	1
affected by fuelwood	ha/yr		700000

For notes see page 80

3.11. APPENDIX

IMAGE DEFORESTATION MODULE DATA SHEET

Africa	in 2100 scenario to			
	A	B	C	D
remaining closed forest	to be calculated			
remaining open forest	to be calculated			
fallow tot. AF closed	$(+1.4\%/a)$			
swidden cycle closed	20	15	10	5
fallow tot. AF open				
fallow period open				
population function[1]	21525	0.0073	0.0410	2935
percentage rural	80	70	60	50
swidden farmers closed	3.5	2.5	2.0	1.0
swidden farmers open	10	8	6	4
number in family	6	7	8	9
cleared per family	1.3	1.1	0.9	0.7
human area (0.05 ha/cap)	proportional to population growth			
yield tot. cereals sc. A[1]	306	0.087	0.018	3500
yield tot. cereals sc. B[1]	297	0.119	0.020	2500
yield tot. cereals sc. C[1]	241	0.161	0.029	1500
yield tot. cereals sc. D[1]	153	0.135	0.047	1150
consumption food cereals	170	150	120	110
percentage food (vs. feed)	65	75	85	95
area non-cereals	80	70	60	50
self-sufficiency ratio[2]	90	80	70	60
forest plantations sc. A[1]	15.1	1.8E-4	0.06	
forest plantations sc. B[1]	2.72	1.6E-5	0.08	
forest plantations sc. C[1]	0.50	1.9E-6	0.10	
forest plantations sc. D[1]	0.09	2.7E-7	0.12	
cattle number scen. A[1]	59100	0.118	0.014	500000
cattle number scen. B[1]	55000	0.157	0.017	350000
cattle number scen. C[1]	58130	0.233	0.020	250000
cattle number scen. D[1]	55700	0.279	0.025	200000
cattle productivity	4	3.5	3.0	2.5
comm. wood prod. index[1,4] A	0.2001	0.0801	0.0265	2.50
comm. wood prod. index[1,4] B	0.1572	0.0786	0.0318	2.00
comm. wood prod. index[1,4] C	0.0898	0.0561	0.0424	1.60
comm. wood prod. index[1,4] D	0.0408	0.0292	0.0556	1.40
wood prod. int. index	1.0	1.3	1.6	2.0
affected by fuelwood	prop. to rural population growth			

For notes see page 80

IMAGE DEFORESTATION MODULE DATA SHEET

Latin America		in 2100 scenario to	
	unit	1900	1985
remaining closed forest	1000 ha	873000	693155
remaining open forest	1000 ha	307000	239520
fallow tot. AM closed	1000 ha		108612
swidden cycle closed	years	10	10
fallow tot. AM open	1000 ha		61650
swidden cycle open	years	15	15
population function[1]	mln		406
percentage rural	%	40	40
swidden farmers closed	%	9	9
swidden farmers open	%	8	8
number in family	cap/fm	7	7
cleared per family	ha/yr	1.0	1.0
human area (.05ha/cap)	1000 ha		20300
yield tot. cereals sc. A^1	kg/ha		1800
yield tot. cereals sc. B^1	kg/ha		1800
yield tot. cereals sc. C^1	kg/ha		1800
yield tot. cereals sc. D^1	kg/ha		1800
consumption food cereals	kg/cap	140	140
percentage food (vs. feed)	%	95	60
area non-cereals	%	50	50
irrigated area	1000 ha		
self-sufficiency ratio[2]	%	125	90
forest plantations sc. A^1	1000 ha	1.5	7303
forest plantations sc. B^1	1000 ha	1.5	7303
forest plantations sc. C^1	1000 ha	1.5	7303
forest plantations sc. D^1	1000 ha	1.5	7303
cattle number scen. A^1	1000	100000	250000
cattle number scen. B^1	1000	100000	250000
cattle number scen. C^1	1000	100000	250000
cattle number scen. D^1	1000	100000	250000
cattle productivity	ha/hd	4.4	2.4
other deforestation	1000 ha		1000
comm. wood prod. index[1,4] A	ha/yr	48400	2003000
comm. wood prod. index[1,4] B	ha/yr	48400	2003000
comm. wood prod. index[1,4] C	ha/yr	48400	2003000
comm. wood prod. index[1,4] D	ha/yr	48400	2003000
wood prod. int. index		0.03	1
affected by fuelwood	ha/yr		300000

For notes see page 80

3.11. APPENDIX

IMAGE DEFORESTATION MODULE DATA SHEET

Latin America	in 2100 scenario to			
	A	B	C	D
remaining closed forest	to be calculated			
remaining open forest	to be calculated			
fallow tot. AM closed	$(+1.1\%/a)$			
swidden cycle closed	200	15	10	5
fallow tot. AM open				
swidden cycle open				
population function[1]	31836	0.025	0.0346	1259
percentage rural	60	55	50	45
swidden farmers closed	8.0	6.0	4.0	3.0
swidden farmers open	8	6	4	2
number in family	7	7	7	7
cleared per family	1.3	1.1	1.0	0.8
human area $(.05 ha/cap)$	proportional to population growth			
yield tot. cereals sc. A[1]	628	0.150	0.019	4200
yield tot. cereals sc. B[1]	546	0.162	0.023	3400
yield tot. cereals sc. C[1]	394	0.141	0.030	2800
yield tot. cereals sc. D[1]	355	0.158	0.038	2250
consumption food cereals	200	170	150	130
percentage food (vs. feed)	50	60	70	80
area non-cereals	75	65	55	45
irrigated area				
self-sufficiency ratio[2]	115	105	90	85
forest plantations sc. A[1]	49.2	6.3E-4	0.06	79000
forest plantations sc. B[1]	8.54	5.5E-5	0.08	155000
forest plantations sc. C[1]	1.53	6.6E-6	0.10	235000
forest plantations sc. D[1]	0.28	9.0E-7	0.12	310000
cattle number scen. A[1]	52000	0.041	0.021	1270000
cattle number scen. B[1]	53600	0.071	0.023	755000
cattle number scen. C[1]	52060	0.116	0.028	450000
cattle number scen. D[1]	58300	0.167	0.032	350000
cattle productivity	3.5	3.25	3.0	2.5
other deforestation	750	500	250	* 0
comm. wood prod. index[1,4] A	0.0770	0.0193	0.0356	4.00
comm. wood prod. index[1,4] B	0.0607	0.0202	0.0401	3.00
comm. wood prod. index[1,4] C	0.0278	0.0139	0.0535	2.00
comm. wood prod. index[1,4] D	0.0049	0.0033	0.0802	1.50
wood prod. int. index	1.0	1.3	1.6	2.0
affected by fuelwood	prop. to rural population growth			

For notes see page 80

IMAGE DEFORESTATION MODULE DATA SHEET

Tropical Asia		in 2100 scenario to	
13 countries[1]	unit	1900	1985
remaining closed forest	1000 ha	381000	301344
remaining open forest	1000 ha	85000	30653
closed 9 countries	%		92
open 9 countries	%		98
fallow tot. AS closed	1000 ha		69225
swidden cycle closed	years	10	10
fallow tot. AS open	1000 ha		3990
swidden cycle open	years	9	9
population function[2]	mln		1274
percentage rural	%	50	50
% swidden farmers closed	%	2	2
% swidden farmers open	%	0.4	0.4
number in family	cap/fm	9	9
cleared per family	ha/yr	0.7	0.7
human area ($0.05 ha/cap$)	1000 ha		63700
yield tot. cereals sc. A^2	kg/ha		2500
yield tot. cereals sc. B^2	kg/ha		2500
yield tot. cereals sc. C^2	kg/ha		2500
yield tot. cereals sc. D^2	kg/ha		2500
consumption food cereals	kg/cap	170	170
percentage food (vs. feed)	%	95	80
area non-cereals	%	50	50
irrigated area	1000 ha		62000
self-sufficiency ratio[3]	%	100	95
forest plantations sc. A^2	1000 ha	2	7320
forest plantations sc. B^2	1000 ha	2	7320
forest plantations sc. C^2	1000 ha	2	7320
forest plantations sc. D^2	1000 ha	2	7320
cattle number scen. A^2	1000	135000	303000
cattle number scen. B^2	1000	135000	303000
cattle number scen. C^2	1000	135000	303000
cattle number scen. D^2	1000	135000	303000
cattle productivity	ha/hd	0.10	0.10
comm. wood prod. index[2,5] A	ha/yr	68160	1755000
comm. wood prod. index[2,5] B	ha/yr	68160	1755000
comm. wood prod. index[2,5] C	ha/yr	68160	1755000
comm. wood prod. index[2,5] D	ha/yr	68160	1755000
wood prod. int. index		0.04	1
affected by fuelwood	ha/yr		1700000

For notes see page 80

3.11. APPENDIX

IMAGE DEFORESTATION MODULE DATA SHEET

Tropical Asia	in 2100 scenario to			
13 countries[1]	A	B	C	D
remaining closed forest	to be calculated			
remaining open forest	to be calculated			
closed 9 countries	to be calculated			
open 9 countries	to be calculated			
fallow tot. AS closed	(+1.25%/a)			
swidden cycle closed	20	15	10	5
fallow tot. AS open				
swidden cycle open				
population function[2]	138988	0.0433	0.0320	3206
percentage rural	70	65	55	50
% swidden farmers closed	4.0	3.0	2.0	1.0
% swidden farmers open	0.4	0.3	0.2	0.1
number in family	7	7	8	8
cleared per family	1.0	0.8	0.7	0.6
human area ($0.05 ha/cap$)	proportional to population growth			
yield tot. cereals sc. A^2	563	0.095	0.024	6000
yield tot. cereals sc. B^2	348	0.088	0.035	4000
yield tot. cereals sc. C^2	279	0.081	0.041	3500
yield tot. cereals sc. D^2	226	0.075	0.049	3000
consumption food cereals	230	210	190	170
percentage food (vs. feed)	65	75	85	95
area non-cereals	70	60	50	40
irrigated area	150000	125000	100000	85000
self-sufficiency ratio[3]	110	100	90	80
forest plantations sc. A^2	62.9	2.5E-3	0.06	25160
forest plantations sc. B^2	9.53	1.9E-4	0.08	50200
forest plantations sc. C^2	1.65	2.2E-5	0.10	75000
forest plantations sc. D^2	0.29	2.9E-6	0.12	100000
cattle number scen. A^2	150000	0.040	8.6E-3	3750000
cattle number scen. B^2	148360	0.099	0.011	1500000
cattle number scen. C^2	109970	0.220	0.023	500000
cattle number scen. D^2	4462	0.013	0.075	350000
cattle productivity	0.30	0.25	0.15	0.10
comm. wood prod. index[2,5] A	0.1846	0.1055	0.0317	1.75
comm. wood prod. index[2,5] B	0.1150	0.0767	0.0408	1.50
comm. wood prod. index[2,5] C	0.0390	0.0300	0.0589	1.30
comm. wood prod. index[2,5] D	1.0E-4	1.0E-4	0.1495	1.00
wood prod. int. index	1.0	1.3	1.6	2.0
affected by fuelwood	prop. to rural population growth			

For notes see page 80

Notes concerning Africa:

1 A,B,C in formula pop $= A/\{B + \exp(-Ct)\}$, D: saturation level; base year 1900
2 selfsufficiency ratio = consumption/production (when < 100: import/food aid)
3 agricultural expansion into non-forest areas
4 wood production index is 1 in 1985

FAO: in 1983 forest+woodlands 719.6 mln. $ha.$, pasture 778.2 mln. $ha.$, agricultural 183.1 mln. $ha.$ and other lands 1277.1 mln. $ha.$

Notes concerning Latin America:

1 A,B,C in formula pop $= A/\{B + \exp(-Ct)\}$, D: saturation level; base year 1900
2 self-sufficiency ratio = consumption/production (when < 100: import/food aid)
3 agricultural expansion into non-forest areas
4 wood production index is 1 in 1985

FAO: in 1983 forest+woodlands 932.7 mln. ha, agric. land 175.5 mln. ha, pasture 550.4 mln ha and other lands 291.6 mln. ha

Notes concerning Tropical Asia:

1 Indonesia, Papua New Guinea, Burma, Malaysia, Philippines, Thailand, Vietnam, Kampuchea, Laos, Bhutan, Nepal, India, Bangladesh
2 A,B,C in formula pop $= A/\{B + exp(-C*t)\}$, D: saturation level; base year 1900
3 self-sufficiency ratio = consumption/production (when < 100: import/food aid)
4 agricultural expansion into non-forest areas
5 wood production index is 1 in 1985

FAO: in 1980 forest+woodlands 336.5 mln. $ha.$, pasture 30.2 mln. $ha.$, agricultural land 257.2 mln. $ha.$ and other lands 185.0 mln. ha

Chapter 4

The Methane Module

4.1 Introduction

Methane (CH_4), the most abundant hydrocarbon in the atmosphere, is considered as an important greenhouse gas, second only to carbon dioxide (Ramanathan et al., 1985, Wang and Molnar, 1985). Analysis of air contained in ice cores indicates that the methane concentration has been ca. 0.7 *ppm* for maybe thousands of years (Khalil and Rasmussen, 1982, and 1987). Around 1700 A.D. the concentration seems to have started increasing slowly to ca. 0.9 *ppm* around 1900 A.D. Only then did a rapid increase start, probably caused by changes in the sources as well as in the sinks of methane resulting in a present concentration of about 1.7 *ppm* (Khalil and Rasmussen, 1987). The anthropogenic sources of methane, such as cattle breeding, rice cultivation and the exploitation of fossil fuels have grown over the last centuries, whereas the natural emissions are assumed to have remained unchanged (Khalil and Rasmussen, 1982, and 1985).

The main sink of methane is oxidation by hydroxyl (OH) radicals. Besides this sink methane is removed from the troposphere by soils and transportation to the stratosphere. Due to the increased levels of methane and carbon monoxide (CO), the dominant consumers of hydroxyl, the hydroxyl concentration has decreased. Thus the removal rate of methane has decreased (the atmospheric lifetime of methane has increased). This combination of growing emissions and declining removal rate has caused the increase in the concentration of methane (Sze, 1977, Bolle et al., 1986, Crutzen, 1985, Khalil and Rasmussen, 1985, Isaksen and Høv, 1987).

This study describes the global CH_4-CO-OH cycle by both quantifying future emissions and simulating the main atmospheric chemical processes that influence the global concentrations of these trace gases. Moreover, the resulting temperature increase is calculated. This has resulted in the implementation of an independent, separate methane module within IMAGE (Rotmans et al., 1990b).

The purpose of the present study is to gain more insight into the driving forces of the CH_4-CO-OH cycle as well as to present estimates of possible future changes in chemical composition and temperatures of the atmosphere. As opposed to previous studies (Brühl and Crutzen, 1988, Thompson and Cicerone, 1986, Isaksen and Høv, 1987), this study includes an extensive investigation of the sources of methane and carbon monoxide, leading to integrated future emission scenarios based on social and economic trends.

4.2 Model Description

4.2.1 Structure

The developed simulation model, an upgraded version of the methane module in Rotmans (1986), is time-dependent with a simulation time of 200 years, from 1900 to 2100. In view of the time scale of our simulation model, the system has a global character, entailing absence of spatial dimensions. The structure of the methane module is depicted in Figure 4.1, which figure shows the cyclic character of the system.

4.2.2 Calculation Procedure

At an arbitrary point of time $T = t$ during the simulation, emissions and concentrations are supposed to be known, either by initialization or by calculation in a previous time step. Methane is removed from the troposphere by three processes, the most important of which is considered first: oxidation by OH radicals. The production of OH radicals is calculated from global trends in tropospheric ozone (O_3) and NO_x. In our model the concentration of OH radicals is determined by the production of OH on the one hand and the loss rate due to reaction with CH_4 and CO on the other hand.

The concentrations of CH_4 and OH determine the rate of removal of CH_4. The two other sinks of CH_4 considered are uptake by soils and transport to the stratosphere, both considered as fixed fractions of the methane concentration at time $T = t$.

4.2. MODEL DESCRIPTION

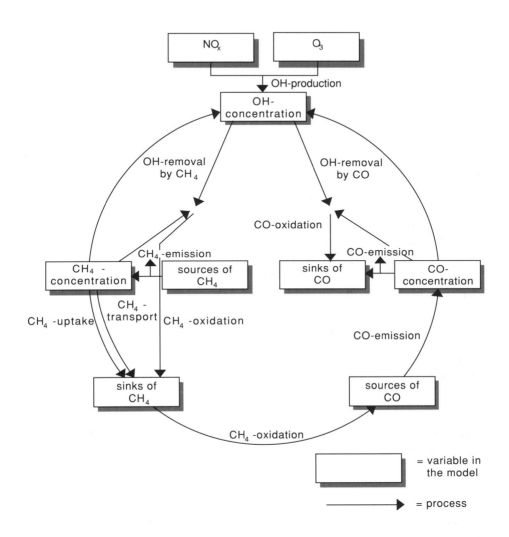

Figure 4.1: Schematic representation of IMAGE CH_4-CO-OH module.

At time $T = t + \Delta t$ the CH_4 concentration is simulated by adding to the CH_4 concentration at time $T = t$ the CH_4 emissions during the interval Δt and subtracting the removal from the troposphere by the three processes mentioned.

The main removal processes of CO are the oxidation by OH as well as the uptake by soils. Both are modelled identically to CH_4: the oxidation by calculation from the concentrations of CO and OH, and the uptake as a fixed fraction of the CO concentration.

At time $T = t + \Delta t$ the CO concentration is calculated by adding the CO emissions during the interval Δt and subtracting the removal of CO during Δt.

Obviously the above mentioned processes are highly interdependent and therefore the CH_4-CO-OH system can be considered as a cycle. In the following sections this system is described in more detail.

4.2.3 Notation

Henceforth the following conventions for notation hold :

$pX(t)$ = the tropospheric concentration of a trace gas at time t (in ppm)
$pX(0)$ = the initial tropospheric concentration of a trace gas at time $t = 0$; here, as $t = 0$, 1900 is selected
$emX(t)$ = the global emission of a trace gas at time t (in Tg/y)
k_1 = the reaction rate constant for the reaction of CH_4 and OH
k_2 = the reaction rate constant for the reaction of CO and OH

4.3 Emissions

Scenarios

For the period 1985 to 2100 four social tendencies are regarded, leading to four sets of emission scenarios. An emission scenario is defined as an integrated perception of a possible development of future emissions, without pretending to be a prediction. Furthermore, we define a set of scenarios as a similar, consistent development for all trace gases. The underlying scenario assumptions are based on trends in energy supply, agriculture, industry, world population growth, etc., as illustrated in Rotmans et al. (1990a). The four sets of scenarios are characterized as follows:

- scenario **A**: *continued trends* assumes a continuation of economic growth not limited by environmental constraints
- scenario **B**: *reduced trends* are based on the presently considered environmental measures
- scenario **C**: *changing trends* assumes implementation of international movement towards stricter policies

4.3. EMISSIONS

- scenario **D**: *forced changes* assesses the possibilities of maximum efforts towards global sustainable development.

Methane

The sources of methane all emit directly into the atmosphere; no methane is formed in the atmosphere itself. The sources are divided into anthropogenic and natural sources. Anthropogenic sources are all controlled or influenced by man, such as rice paddies, cattle, biomass burning, fossil fuel usage and leakage of natural gas. The major natural sources are termites, swamps, fresh- and salt water and tundras (Seiler, 1984, Bartlett et al., 1985, Ehhalt, 1985, Crutzen et al., 1986a, Bingemer and Crutzen, 1987, Van Ham, 1987). In general, methane emissions inventories agree upon the nature of the sources, but the lower and the upper bound for the total emission differ by a factor of 6 (Van Ham, 1987, Bingemer and Crutzen, 1987). The possibility that the fossil carbon sources of methane could be more important than is commonly believed (Lowe et al., 1988) has not yet been taken into account. Table 4.1. gives a summary of the various sources and the corresponding methane emissions.

sources	emissions (in $TG\ y^{-1}$)	range in (in $TG\ y^{-1}$)
total	490	213 – 1397
anthropogenic	325	190 – 720
rice paddies	70 (b)	30 – 200
ruminants	75	60 – 200
biomass burning	60 (c)	20 – 110
natural gas leakage	30 (c)	20 – 40
coal mining	30 (c)	30 – 100
waste disposals (a)	60	30 – 70
natural	165	23 – 677
termites	165	6 – 300
swamps/wetlands	100 (c)	25 – 300
fresh and salt water	5	0 – 12

From: Van Ham 1987
except   (a) : from Bingemer and Crutzen 1987
             (b) : also from Holzapfel-Pschorn and Seiler 1986
             (c) : also from Crutzen 1985

Table 4.1: Estimated emissions of methane sources in 1985.

Historical emissions of methane for the years 1900 to 1985 have been estimated in Bolle et al. (1986) and Khalil and Rasmussen (1985) and follow an exponential growth function.

Four scenarios have been developed for presumed future emissions for the period 1985 to 2500. These scenarios are derived from the four sets of scenarios, which are, together with the underlying assumptions for the CH_4, the emission scenarios described in Rotmans et al. (1990a). In Figure 4.2 an estimate of the build-up of the methane emissions in 2050 is given, disregarding the temperature feedback mechanisms on the methane emissions. Methane is a useful gas and therefore at some sources it is attractive to recover it, not only out of environmental concern, but also for economic reasons. This is particularly true for emissions during the exploitation of fossil fuels and from wastes. From scenario A to D decreasing estimates for the source strengths are used, as well as an increasing recovery rate for the relevant sources.

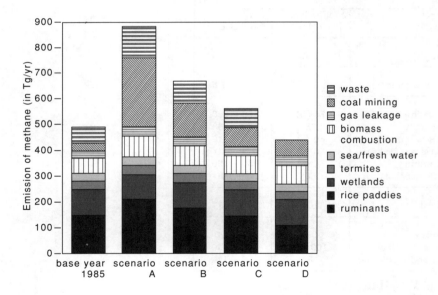

Figure 4.2: Build-up of the methane emissions in 2050 (from R.J. Swart, 1988).

4.3. EMISSIONS

Carbon monoxide

Carbon monoxide is both emitted directly as well as produced in the atmosphere by oxidation of hydrocarbons, of both natural and anthropogenic origin. The direct sources are mainly the burning of biomass and fossil fuel, plants and oceans.

We distinguish three kinds of carbon monoxide emissions (in fact what we call emission is in-situ production):

- from the oxidation of CH_4;
- from the oxidation of hydrocarbons other than CH_4;
- from the direct sources.

In the model these specifications are expressed in the following way:

$$EmCO(t) = EmCO_{CH_4}(t) + EmCO_{NMHC}(t) + EmCO_{direct}(t) \quad (4.1)$$

with:
$EmCO(t)$ = the total CO emissions at time t (in Tgy^{-1})
$EmCO_{CH_4}(t)$ = the CO emissions at time t due to CH_4 oxidation (in Tgy^{-1})
$EmCO_{NMHC}(t)$ = the CO emissions at time t due to the oxidation of nonmethane hydrocarbons (NMHC) (in Tgy^{-1})
$EmCO_{direct}(t)$ = the CO emissions at time t form the direct sources (in Tgy^{-1})

The CO emission due to CH_4 oxidation is modelled by the net CO yield from the methane oxidation by hydroxyl, which is approximately 80% according to Logan et al. (1981). This gives a source strength of 80% of the product of the CH_4 concentration and the oxidation velocity; see Rotmans and Eggink (1988).

In order to quantify the direct emissions of CO and the oxidation by NMHC the relevant sources of these gases are divided into three categories: natural, fossil fuel related and other anthropogenic emissions. While for natural emissions of CO and NMHC constant figures were taken from Logan et al. (1981), for anthropogenic but non-fossil fuel related emissions growth has been assumed proportional to population. Sources treated this way are carbon monoxide and hydrocarbon emissions from deforestation, agricultural practices like savanna burning, solvent use, as well as residential combustion of fossil fuel. Thompson and Cicerone (1986) assumed all fossil fuel related CO emissions and the oxidation of CO by anthropogenic hydrocarbons to be proportional to fossil fuel consumption.

However, the emissions of these trace gases are the result of incomplete combustion of fossil fuels (especially in traffic, residential heating) or losses during storage, transportation or use of non-energy fuel products (NMHC by evaporation). In future energy scenarios centralized fossil fuel combustion in the industrial and the electricity generating sector, where carbon monoxide and hydrocarbon emissions are low, plays a dominant role. Therefore this assumption appears incorrect. In this study alternative scenarios are adopted (Swart, 1988). Since ca. 80% of the global fossil fuel related CO emissions and ca. 50% of the NMHC emissions are presently caused by vehicular traffic, scenarios for this sector have been created, based on car densities.

Following an OECD methodology (OECD, 1983), saturation values of cars have been used for seven different regions consisting of countries with assumed similar characteristics as to car ownership. Extrapolating historical figures (Automobile International, 1929–1987) and using four sets of saturation values, the numbers of cars, trucks and buses have been calculated, using the mean UN-population scenario (UN, 1986) and logistical curves for both population and car ownership. The results are shown in Figure 4.3. With presently available technology, like catalysts and other engine types, emissions of CO and NMHC can be greatly reduced (up to 90%) and fuel efficiency increased (factor 3, Cheng et al., 1986, Goldenberg et al., 1987a,b).

In the CO scenarios it is assumed that, from A to D, the emission reduction percentages and fuel efficiency are increasing and have been introduced earlier. In developing countries the reduction percentages are assumed to be lower for reasons of worse maintenance and inapplicability of technologically sophisticated features and also introduced later, allowing for a time lag in law making and enforcement. The base year estimate for CO emissions from traffic is taken from Kavanaugh (1987), whose 152 Tg/yr appears to be more realistic than Logan's (1981) 233 Tg/yr, according to Swart (1988). Another major CO source is the primary metal industry. Not only can the emissions be reduced by process changes or improvements, remaining CO can also be flared or reused. While iron and steel production per capita is already decreasing and aluminium production approaching saturation value, in the scenarios the unabated global production levels are assumed to be proportional to population growth, like other anthropogenic emissions of CO and NMHC, apart from traffic. Even if high car ownership levels are combined with minimal emission reduction, the resulting emission levels are still much lower than would be the case with proportionality to fossil fuel consumption.

4.3. EMISSIONS

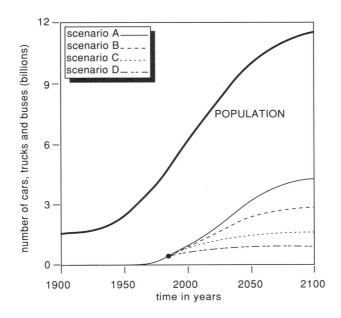

Figure 4.3: Number of cars, trucks and buses (from R.J. Swart, 1988).

Hydroxyl

The main source of hydroxyl radicals is the reaction of exited oxygen $(O(^1D))$, produced by the photolysis of ozone (O_3), and water (H_2O) (Hewitt and Harrison, 1985, Logan et al., 1981). Other (minor) sources are the recycling reactions of odd hydrogen (OH, H, HO_2, H_2O_2) (Logan et al., 1981, Sze, 1977).

The limiting condition at the production of OH radicals is the magnitude of the NO_x concentration. Hameed et al. (1979) show the dependence of the CH_4-CO-OH cycle on the NO_x background concentration.

Here a low ambient NO_x concentration is assumed which involves a decreasing OH concentration in case of increasing CO and CH_4 concentration. The OH production is presumed to be commensurate with the concentrations of NO_x and O_3 (Logan et al., 1981). A study of the literature shows general agreement on the fact that there has been an increase in tropospheric ozone and NO_x in the last century (Ramanathan et al., 1985, Crutzen, 1985, Thompson and Cicerone, 1986, Brasseur and de Rudder, 1987). Therefore, a rising historical OH production is implemented, under the assumption of a steady-state OH concentration and calibrated on the OH concentration; this yields an increase from ca. 0.3 $ppm\ y^{-1}$ in 1900 to 0.6 $ppm\ y^{-1}$ in 1985. For the years 1986 to 2100 four OH production scenarios have been chosen, based on presumed future developments in tropospheric O_3 and NO_x concentrations, according to Stewart et al. (1977), Brühl and Crutzen, (1985), Crutzen and Graedel, (1986b) and Isaksen and Høv, (1987).

Annual increases of both NO_x and O_3 concentrations are varied from +0.40% (scenario A), via +0.20% (scenario B) and +0.10% (scenario C) to +0.05% (scenario D). Increases of abatement measures to combat acid rain have not been explicitly taken into account.

The production of OH is roughly determined by the sum of the trends in NO_x and O_3, neglecting the other minor sources of hydroxyl radicals (Logan et al., 1981). The applied scenarios do not yet take account of the actual sources for lack of more detailed information.

4.4 Concentrations

Methane

The tropospheric concentration of methane is determined by the initial concentration, the emissions of methane and its removal. Three removal processes are considered: the uptake of methane by soils, the transportation of methane to the stratosphere and the oxidation of methane. The rates of the three processes considered here determine the atmospheric lifetime. We assume the uptake velocity by soils and the transportation velocity to the stratosphere to be time-independent and the oxidation velocity to be time-dependent. This leads to the following expression for the CH_4 concentration:

$$pCH_4(t) = pCH_4(t-1) + \int_{t-1}^{t} \quad \text{(conversion factor } CH_4 * EmCH_4(\tau)$$

$$- \text{ uptake velocity } CH_4 * pCH_4(\tau)$$
$$- \text{ transport velocity } CH_4 * pCH_4(\tau)$$
$$- \text{ oxidation velocity } CH_4 * pCH_4(\tau))d\tau$$

(4.2)

with:

$pCH_4(0)$ = the initial CH_4 concentration, at time $t=0$ (in 1900); is 0.9 ppm, varying between 0.85 ppm (Khalil and Rasmussen, 1985, 1987) and 1.1 ppm

4.4. CONCENTRATIONS

conversion factor CH_4 = (Kerr, 1984) factor that converts emissions of CH_4 into global tropospheric concentrations of CH_4; is $3.76 * 10^{-4}$ (in $ppm\ Tg^{-1}$) (Khalil and Rasmussen, 1982)

uptake velocity CH_4 = the rate with which CH_4 is taken up by soils (in y^{-1}). $0.005\ y^{-1}$, based on Van Ham (1987) and Keller et al. (1986)

transp. velocity CH_4 = the rate with which CH_4 is transported to the stratosphere (in y^{-1}). $0.015 y^{-1}$, according to Van Ham (1987)

oxidat. velocity CH_4 = the rate with which methane is oxidized by OH at time t (in y^{-1}). Following Volz et al. (1981), Logan et al. (1981) and Khalil and Rasmussen (1985) the product of the reaction constant k and the OH concentration

A derived concept is the atmospheric life time of methane, defined as:

$$\begin{aligned} \text{lifetime } CH_4(t) &= 1/(\text{ uptake velocity } + \text{ transp. velocity} \\ &\quad + \text{ ox. velocity}) \\ &= 1/(0.005 + 0.015 + k_1 * pOH(t)) \end{aligned} \quad (4.3)$$

Carbon monoxide

Similar to methane, the CO concentration is determined by the initial concentration, the emissions and the removal from the troposphere. The major removal mechanism for CO is the oxidation by OH radicals, which accounts for ca. 80% of the CO sink. A second important sink is the uptake by soils (Logan et al., 1981, Volz et al., 1981, Khalil and Rasmussen, 1984, and 1985, Thompson and Cicerone, 1986). Although there may be some CO transport to the stratosphere or CO uptake by plants (Khalil and Rasmussen, 1984), these contributions are neglected in the model. The CO concentration is expressed as:

$$pCO(t) = pCO(t-1) + \int_{t-1}^{t} (\text{conversion factor } CO * EmCO(\tau) \\ - \text{ uptake velocity } CO * pCO(\tau) \\ - \text{ oxidation velocity } CO * pCO(\tau))d\tau \quad (4.4)$$

with:

$pCO(0)$	= 0.048, varying between 0.045 *ppm* and 0.085 *ppm* for 1860 (Thompson and Cicerone, 1986). We obtained the value of 0.048 *ppm* from a concentration of 0.1 *ppm* for 1977 and an increase of 30% since 1950 (Sze, 1977), assuming an exponentially growing concentration
conversion factor CO	= factor that converts emissions of CO into global tropospheric concentrations of CO; is $2.0186 * 10$ *ppm* Tg^{-1} (Rotmans, and Eggink, 1988)
uptake velocity CO	= the rate with which CO is taken up by soils (in y^{-1}). The uptake of CO by soils is ca. 320 Tgy^{-1} for around 1985 and depends linearly on the CO concentration, argued by Volz et al. (1981), resulting in a time-independent uptake velocity for CO of 0.64595 y^{-1}
oxidation velocity CO	= the rate with which CO is oxidized by OH at time t (in y^{-1}). Following the interpretation of Logan et al. (1981) and Volz et al. (1981), this is the product of a reaction rate constant and the OH concentration

Hydroxyl radicals

The change in hydroxyl concentration is determined by the production (already discussed in the section emissions) and the loss of hydroxyl radicals. It is generally accepted that hydroxyl is the dominant consumer of methane and carbon monoxide, and conversely, methane and carbon monoxide are the major sinks of hydroxyl (Hewitt and Harrison, 1985, Logan et al., 1981). Therefore the reaction rates for the reactions of CH$_4$ and CO with OH determine the loss rate of OH, which, together with the hydroxyl concentration, determines the loss of OH :

$$\text{loss rate } OH(t) = k_1 * pCH_4(t) + k_2 * pCO(t) \tag{4.5}$$

with:

loss rate $OH(t)$	= the removal rate of OH from the troposphere at time t by the reactions with CH$_4$ and CO.
k_1	= $1.584 * 10$ ppm^{-1} y^{-1}, see Rotmans and Eggink, (1988)

k_2 = $8.91 * 10$ ppm^{-1} y^{-1}, from $1.35 * 10(1 + p/1000)$ cm $molec^{-1}s^{-1}$ with p = atmospheric pressure in mbar, see Rotmans and Eggink, (1988)

The foregoing results in a loss rate of ca. $10^{-7}y^{-1}$, which means a tropospheric lifetime of hydroxyl of ca. 3 seconds. This implies that the loss of OH approximately equals the production of OH, justifying the supposition of a steady-state concentration of hydroxyl, or a negligible annual change in OH concentration, so:

$$pOH(t) = \text{production } OH(t)/(k_1 * pCH_4(t) + k_2 * pCO(t)) \qquad (4.6)$$

based on Logan et al. (1981) and Levine et al. (1985).

4.5 Results

The emissions of CO_2 (from fossil fuels), CH_4 and CO are depicted in figures 4.4–4.6. These figures show a sharp rise of the emissions in scenario A (*continued trends*). Scenarios B and C (*reduced* and *changing trends*) lead to logistical emission curves (showing a deflection owing to environmental measures); scenario D (*forced changes*) results in a substantial decrease of the CO_2 emission, a slight fall of the CH_4 emission and a slightly increasing CO emission.

Figures 4.4 and 4.6 clearly show the absence of a correlation (as supposed in literature) between the fossil fuel CO_2 emission and the CO emission.

Figure 4.7 shows the OH production, ranging between a large increase of the production of hydroxyl radicals until 2100 for scenario A and a slightly increasing production until 2100 for other scenarios.

The calculated global tropospheric concentrations of CH_4, CO and OH are represented in Figures 4.8–4.10. Figure 4.8 shows an upward trend for scenarios A and B with a deflecting curve after 2050, a relatively constant CH_4 concentration for scenario C; only scenario D leads to a slightly decreasing CH_4 concentration.

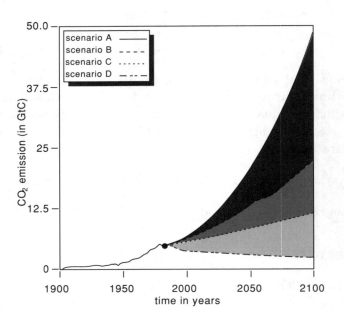

Figure 4.4: Emissions of CO_2.

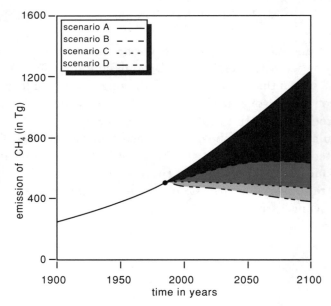

Figure 4.5: Emissions of CH_4.

4.5. RESULTS

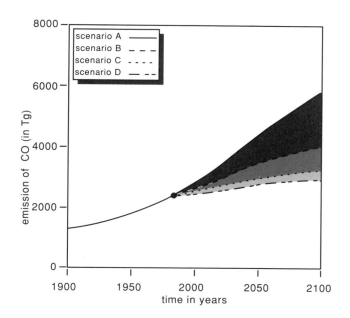

Figure 4.6: Emissions of CO.

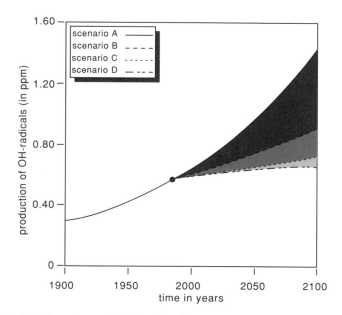

Figure 4.7: Production of OH radicals.

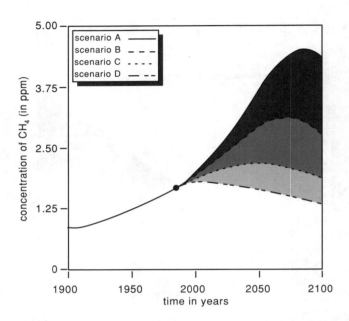

Figure 4.8: Concentration of CH_4 (global tropospheric average).

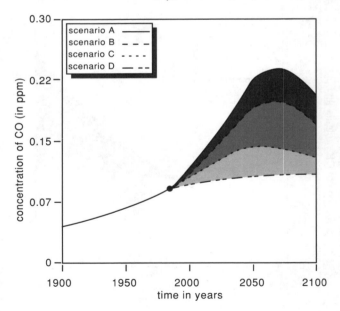

Figure 4.9: Concentration of CO.

4.5. RESULTS

Figure 4.9 shows a growth of the carbon monoxide concentration in case of scenarios A, B and C, stagnating after 2075; and even scenario D induces a minimal growth of the CO concentration. Both the CH_4 concentrations and the CO concentrations show logistical behaviour. Primarily this is caused by the deflecting emissions, with a turning point about 2050; the system's cyclic character strengthens also this effect: decreasing CO concentrations have a restraining influence on the CH_4 concentrations.

A remarkable result is that in our model even major environmental measures and maximum efforts towards global sustainable development cannot effectuate a robust falling-off of the concentrations. This is due to remaining methane emissions and those emissions of carbon monoxide, which are taken as being proportional to population growth.

In Figure 4.10 the calculated OH concentration is depicted, showing for scenario A,B and C a decreasing trend until 2050, followed by an upward trend. As mentioned before this is connected with abating CO concentrations after 2050. The strong growth of the OH concentration in scenario A is due to both decreasing CO and CH_4 and the relatively high OH production. Only scenario D yields a steady growth of the concentration.

The simulated atmospheric lifetime of CH_4 is given in Figure 4.11. In 1900 the residence time of methane in the atmosphere is ca. 9 years. The lifetime path of methane is inversely proportional to the OH concentration (equation 4.3). Scenario D yields a slightly decreasing lifetime of CH_4.

Feedbacks

Two major feedbacks which potentially may increase methane concentrations as a result from global warming have been identified. Methane hydrates have a large potential for being released at increasing rates when global temperatures increase. Lashof (1989) argues that this is the largest biogeochemical feedback in the climate system. According to Kvenvolden (1989) however, this increase is not yet expected to materialize in the coming decades. Another important potential feedback is formed by the increased decomposition of organic materials in anaerobic sediments in northern wetlands (Lashof, 1989). This additional release is modelled in a very provisional way by Burke et al. (1990) by assuming a simple relationship between methanogenesis and temperature. Next to these positive feedbacks there are also negative feedbacks with respect to methane. For instance the decrease in the concentration of methane by increased OH concentration, caused by increase in absolute humidity accompanying global temperature increase.

Current scientific opinion is that the positive feedbacks largely dominate the negative feedbacks.

We have considered the relation between methanogenesis and temperature increase, supposing a 10 degree centigrade temperature increase would cause a fivefold increase of methane production, based on Burke et al. (1990).

This is modelled in a simple way by assuming a linear relationship between methanogenesis and temperature, following Burke et al. (1990):

$$Emwtfb(t) = Emwt(t) * [1 + 0.4 * Ttot(t)] \qquad (4.7)$$

with:
$Emwtfb(t)$ = Emission of wetlands with temperature feedback at time t
$Emwt(t)$ = Emissions of wetlands without feedback at time t
$Ttot(t)$ = total equilibrium temperature increase

According to this relation a 10 degree celsius temperature increase would cause a fivefold increase of methane production. When this relationship is built into IMAGE, assumptions must be made for the developments of the other trace gases. Simulation runs were performed for the unrestricted trends scenario (A). Figure 4.12 shows the effect of the temperature feedback on wetland emissions, whereas Figure 4.13 gives the different CH_4 concentrations with and without feedback.

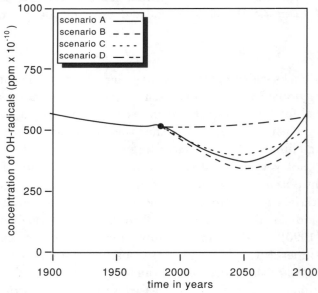

Figure 4.10: Concentration of OH radicals.

4.5. RESULTS

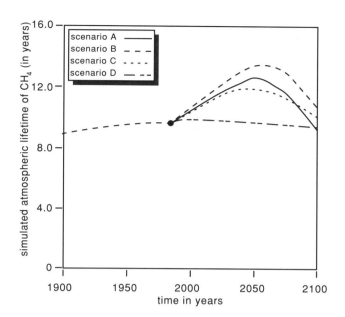

Figure 4.11: Simulated atmospheric lifetime of CH_4.

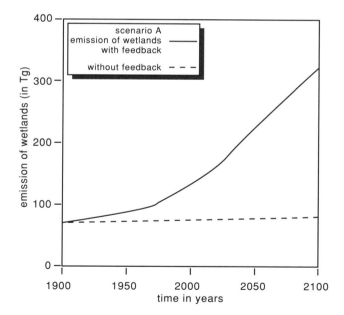

Figure 4.12: CH_4 emission from wetlands with and without climate feedback.

Figure 4.13 shows that methane concentrations might result to be about 50% higher by 2100 if this feedback would materialize.

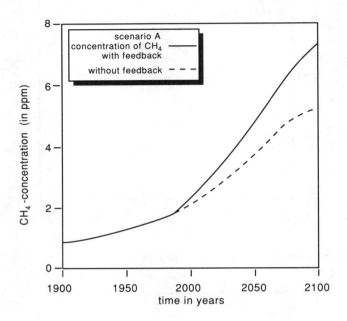

Figure 4.13 CH_4 concentration for the unrestricted trends scenario (A) with and without climate feedback.

4.6 Conclusions

Methane

A considerable part of the sources of *methane* is anthropogenic in nature. At the moment the anthropogenic contribution amounts to approximately 66% and in the *continuing* and *reduced trends* (A and B) it is growing in the coming decades, ranging from 81% to 74% in 2050, mainly due to the increase in coal mining, ruminants, waste disposals and biomass combustion. Only in the *changing trends* and *forced changes* (C and D) does the anthropogenic share gradually decrease to 65% and 63% in 2050, respectively.

Unlike in the case of CO_2 it may be advantageous to limit some anthropogenic methane emissions, methane being a useful gas. Examples are

4.6. CONCLUSIONS

reduction of losses during fossil fuel exploitation, waste disposal and animal husbandry. This is particularly interesting, since the contribution of methane to the greenhouse effect per molecule is greater than that of CO_2 and measures will quickly have an effect.

Even far-reaching environmental measures cannot prevent methane concentrations from rising during the coming century; a doubling of the methane concentration in comparison with the value in 1985 (1.7 *ppm*) can take place as early as the year 2040. It needs no emphasis that only *forced changes* can bring about a real decrease of the methane concentration.

Carbon monoxide, hydrocarbons and nitrogen oxides

Next to methane emissions, emissions of *carbon monoxide* and *non-methane hydrocarbons* strongly influence the methane concentration in the atmosphere. In the international literature the emissions of these gases have been taken to be proportional to global fossil fuel consumption or economic growth. In contrast to CO_2, CO and NMHC are the result of incomplete combustion of fuels or inefficient use of non-energy fuel products. Preliminary scenarios have therefore been developed for this study, which take the sources into account. While CO and NMHC are useful products, it is quite possible, that emissions are reduced by increasing combustion efficiencies, reuse of emitted gas and, for NMHC only, reduction of evaporation losses not only for environmental, but also for economic reasons.

Emissions of carbon monoxide and NMHC from fuelwood combustion, deforestation practices and savanna burning, especially in the third world, are harder to control than fossil fuel related emissions and deserve special attention. Deforestation may not only contribute to the greenhouse effect by direct CO_2 emissions, but also indirectly, via the influence on the CH_4-OH-CO cycle.

Like methane, for carbon monoxide, hydrocarbons and nitrogen oxides, fossil fuel related emissions can be limited by other means than shifts towards non fossil fuels and fuel efficiency increases. Measures presently being introduced or developed to combat acid rain (catalysts, lean burn engines) also contribute indirectly to the prevention or delay of the greenhouse effect.

In the model increasing NO_x emissions have a decreasing influence on CH_4 concentrations via OH. The contribution of NO_x to the greenhouse effect as a precursor of the greenhouse gas ozone is not considered in this study.

General

With the help of the developed simulation model global trends of emissions and concentrations of CH_4, CO, OH as well as the temperature effect of methane can be calculated. The model shows that there is no sense in a forced reduction of only the methane emissions. Even if the emissions were to be stabilized the methane concentrations would still rise, due to the interrelated influence of other trace gases like carbon monoxide and hydroxyl (by NO_x and O_3). The sources of NO_x and CO are especially important.

In the model, in the second half of the 21st century the growth of the concentrations and consequently the temperature effect of CH_4 reverses due to an increase in OH concentration. This increase is caused by a combination of stabilizing CO emissions and continuing OH production. Simulations with the CH_4-CO-OH cycle model, considered in this study, underline the importance of methane as a greenhouse gas.

Finally it is worth mentioning that the CH_4-CO-OH cycle, functioning as an internal feedback system, is a system which is critically sensitive to external perturbations. Further investigation of this system is necessary.

Chapter 5

The N_2O Module

5.1 Introduction

Nitrous oxide or laughing gas (N_2O) is produced both naturally and artificially. This greenhouse gas is produced and consumed by biological processes in soil and water and is removed from the atmosphere much more slowly than methane. The yearly growth in the N_2O emission trend varied from 0.1% at the beginning of this century to 1.3% in the last 10 years. The current average concentration of N_2O is about 0.308 *ppm* and is increasing at about 0.2 to 0.3 percent a year (0.18 − −0.26% according to Lyon et al., 1989). For N_2O a simple one-box emissions and concentration module has been implemented.

5.2 N_2O Emissions Module

We are not yet able to quantify N_2O emissions. Both the anthropogenic and natural sources are subject to large uncertainty. Anthropogenic sources of N_2O are strongly related to energy use (combustion of fossil fuels and biomass burning) and agriculture (nitrogen fertilizers). Although biomass burning and use of nitrogen fertilizers are the dominant sources of the anthropogenic emission of N_2O, other human activities also contribute to the atmospheric N_2O concentration, such as acid rain and forest clearing (Marland and Rotty, 1985, Robertson and Tiedje, 1988). In the emission module three major anthropogenic sources are taken into account: combustion of fossil fuels, biomass burning and fertilization, represented in Equation (5.1) (Marland and Rotty, 1985, van Ham, 1987, Conrad et al., 1980 and 1983,

Crutzen, 1983, Hao et al., 1987, Kavanaugh, 1987, and Lyon et al., 1989).

$$N_2OHUM(t) = N_2OFSE(t) + N_2OBIO(t) + N_2OMAN(t) \qquad (5.1)$$

with:
$N_2OFSE(t)$ = Emissions of N_2O by fossil fuel combustion at time t
$N_2OBIO(t)$ = Emissions of N_2O by biomass burning at time t
$N_2OMAN(t)$ = Emissions of N_2O by manure soils at time t

The natural emissions of N_2O are primarily caused by the nutrient cycling in soils and aqueous systems. Three major natural sources are distinguished in the emission module: forest soils, other soils and oceans (Banin, 1986, Marland and Rotty, 1985, Elshout, 1989, Chamberlain et al., 1982, Keller et al., 1986, Anderson and Levine, 1987), which are given in Equation (5.2):

$$N_2ONAT(t) = N_2OFOR(t) + N_2OSOI(t) + N_2OOCE(t) \qquad (5.2)$$

with:
$N_2OFOR(t)$ = Emissions of N_2O by forests soils at time t
$N_2OSOI(t)$ = Emissions of N_2O by other soils at time t
$N_2OOCE(t)$ = Emissions of N_2O by oceans at time t

There is little consensus as to the quantification of both the natural and anthropogenic sources of N_2O. Recent estimations of N_2O emissions (van Ham, 1987, Lyon et al., 1989, EPA, 1989, Elshout, 1989) are much lower than earlier estimations (Chamberlain et al., 1982, Hahn, 1974, Hahn and Junge, 1977), particularly with respect to the contribution of the oceans and the combustion of fossil fuels. Lyon et al. (1989), and EPA (1989) suggest that, although nothing has been proved beyond the shadow of a doubt, combustion of fossil fuels is not a significant source of N_2O. Additionally the emissions by lightning can be neglected according to Hill et al. (1984), and Levine and Shaw (1983). Table 5.1 surveys the diversity of estimations of N_2O emissions.

5.2. N_2O EMISSIONS MODULE

sources	Hahn and Junge 1977	Khalil and Rasmussen 1983	Crutzen 1983	Bolle et al. 1986	Kavanaugh 1987	van Ham 1987
oceans	16 – 185	9.0	1 – 2	2	2	1 – 10
natural soils	6 – 65	13.4	–	6	7.7	3 – 9
cultivated soils	–	6.6	1 – 3	0.2 – 0.6	–	1 – 3
manure soils	6 – 20	–	< 3	0.6 – 2.3	0.8	0 – 8
biomass burning	–	–	1.2	1 – 2	0.7	1 – 2
fossil fuel burning	1 – 4	–	1.8	2	4.0	1 – 3
lightning	10 – 55	–	–	< 0.1	–	–
Total	39 – 329	29.0	8 – 11	12 – 15	15.2	7 – 35

Table 5.1: Survey of global emissions in TgN/yr of N_2O from literature.

For the base year 1985 the following estimates of the anthropogenic and natural sources of N_2O are incorporated into IMAGE:

	1985	
forest soils	5.0 TgN/yr	
manure soils	1.5 TgN/yr	
other soils	0.6 TgN/yr	
oceans	2.0 TgN/yr	
burning of biomass	1.5 TgN/yr	
burning of fossil fuels	0.5 TgN/yr +	
Total emissions	11.1 TgN/yr	
Anthropogenic part	3.5 TgN/yr	about 32% of total emission
Natural part	7.6 TgN/yr	about 68% of total emission

For N_2O four emissions scenarios have been developed, of which the underlying assumptions are described in detail in Rotmans et al. (1990a). The highest scenario, A, *unrestricted* trends, assumes a large increase in N_2O emission from coal use, no emission control measures, a rapid increase in fertilizer use in the third world up to European levels, an average estimate of anthropogenic sources and a continuing deforestation (emission from burning and a major influence of soil emissions). Scenario B, *reduced* trends, assumes, besides an increase in N_2O emissions from coal consumption, a gradual increase in fertilizer use in the third world and a gradual reduction in deforestation is assumed. Scenario C, *changed* trends, still assumes an increase in N_2O emissions from coal consumption, a slight reduction in emission rates from fossil fuel combustion, a slow increase in fertilizer use in the third world and a major global effort towards halting deforestation. Finally, scenario D, *forced* trends, is characterized by decreasing N_2O emissions due to a shift towards non-fossil energy sources and implementation of N_2O emission control technologies; a global program to introduce proper forest management, stop deforestation and start large scale reforestation and lastly a limited growth in fertilizer use.

5.3 N_2O Concentration Module

Although the ambient concentration of N_2O is only one-thousandth that of CO_2, N_2O is a relatively strong infrared absorber. N_2O is stable in the troposphere and is destroyed in the stratosphere by photodissociation and by reaction with singlet oxygen. The sink of N_2O is the most important source of NO, which plays a major role in regulating the stratospheric ozone concentration. So an increasing N_2O concentration indirectly threatens stratospheric ozone (Bolle et al., 1986, Jackman and Guthrie, 1985). N_2O is removed from the atmosphere slowly, and has an average residence time of about 150 years. Estimations for the lifetime of N_2O vary from 100–170 years, most estimations agreeing upon 150–170 years. Owing to this stability and long residence time N_2O is spread homogeneously all over the troposphere. In 1985 the average tropospheric N_2O mixing ratio was 307.4 *ppb* (van Ham, 1987).

The atmospheric removal process in IMAGE is simulated by an exponentially delayed emission mechanism, described in Goodman (1974), and Rotmans (1986). The functional form of this exponentially delayed mechanism is given in Equation (5.3). Simulations with IMAGE show that the exponentially delayed emission corresponds with the removal process that is used for CFCs; the latter process is inversely proportional to the atmospheric lifetime of N_2O and is reflected by the negative exponential function $e^{-t/NLF}$, where NLF is the atmospheric lifetime of N_2O, determined in the N_2O concentration model as 170 years. Both interpretations of the N_2O

5.4. RESULTS

removal process produce comparable results.

$$EMNDL(t) = EMNDL(t-1) + \int_{t-1}^{t} [(1/LFTN_2O) * (EMN_2O(\tau) \\ -EMNDL(\tau)/LFTN_2O)]d\tau \qquad (5.3)$$

with:
$EMNDL(t)$ = delayed emission of N_2O at time t (in Tg)
$LFTN_2O$ = atmospheric lifetime of N_2O (is 170 year; estimates of atmospheric lifetime of N_2O vary from 120–170 year (Rotmans, 1986))

The atmospheric concentration of nitrous oxide is determined by the initial concentration, the sources and the sinks of nitrous oxide. The sources are enumerated in the emissions model description, whereas the main sink of nitrous oxide is caused predominantly by stratospheric losses, reflected by the exponentially delayed emissions mechanism, described above. This leads to the following expression for the N_2O concentration.

$$PN_2O(t) = PN_2O(t-1) + \int_{t-1}^{t} [CVFN_2O * (EMN_2O(\tau) \\ -EMNDL(\tau))]d\tau \qquad (5.4)$$

with:
$PN_2O(t)$ = concentration (mixing ratio) of N_2O at time t (in ppb)
$PN_2O(0)$ = initial concentration of N_2O, 285 ppb (estimates vary from 280 to 290 ppb (Weiss, 1981 and Ramanathan et al., 1985)
$CVFN_2O$ = conversion factor of N_2O, is 0.2 ppb/TgN according to Bolle et al. (1986);
$EMN_2O(t)$ = emission of N_2O at time t (in TgN/yr)
$EMNDL(t)$ = delayed emissions of N_2O at time t (in TgN/yr)

5.4 Results

In Figure 5.1 a breakdown of the emissions of N_2O for a particular year is given for the different scenarios. Figure 5.2 gives the total emissions of N_2O for the different scenarios. These N_2O emissions scenarios are considerably lower than those described in Rotmans et al., (1990a), due to more recent moderate estimates of fossil fuels combustion and soils emissions.

The global N_2O concentration, resulting from the emissions scenarios of Figure 5.2, is given in Figure 5.3. Due to the long atmospheric lifetime of this gas all emission scenarios lead to an increase of the concentrations. Because of this aspect nitrous oxide appears to become increasingly important.

5.5 Conclusions

Simulation experiments with IMAGE show that in the long run N_2O will become a greenhouse gas of great importance. Even the most optimistic N_2O emissions scenario, with nearly stabilizing emissions till 2100, leads to an increasing N_2O concentration, which in turn will bring about a rise in temperature. Clearly, policy measures can at least affect the pace with which atmospheric N_2O will increase.

In any case, because of the uncertainty about the various sources, both natural and anthropogenic, additional research is needed to improve our knowledge of the sources and the validity of measurement techniques.

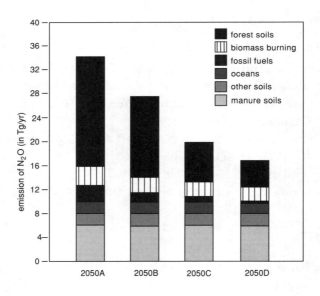

Figure 5.1: Breakdown of the N_2O emissions for the year 2050.

5.5. CONCLUSIONS

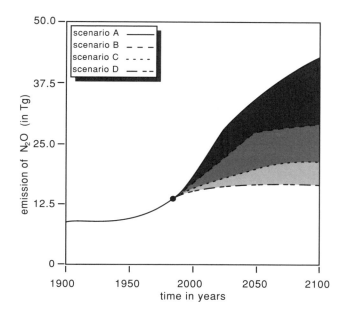

Figure 5.2: Emission of N_2O.

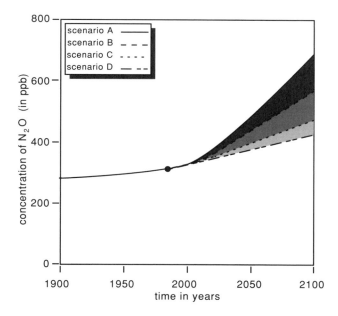

Figure 5.3: Concentration of N_2O.

Chapter 6

The CFCs Module

6.1 Introduction

In the early seventies chlorofluorocarbons, or CFCs, were identified as substances with a strong potential for the depletion of the ozone layer. In the mid eighties the Antarctic ozone hole was discovered, and scientific evidence was found that CFCs, under specific meteorological circumstances, were responsible for this phenomenon. Therefore, in 1987, the Montreal Protocol came into force, implying the regulation of the main chlorofluorocarbons and halons. The regulating mechanism includes an ultimate freeze of 50% in 2000, whereas developing countries are allowed to increase their production, and not all ozone depleting substances are regulated.

The CFCs emissions module of IMAGE provides estimates of emissions of the most important chlorofluorocarbons, CFC-11 and CFC-12. Although dozens of chlorofluorocarbons are currently being produced and emitted to the atmosphere, CFC-11 and CFC-12 are the dominant species.

Recently the module has been extended with other CFCs (especially CFC-113, CFC-114 and CFC-115), halons, methylchloroform and carbontetrachloride (Den Elzen et al., 1990b and c). Future production of CFCs is triggered by policy assumptions rather than economic developments and demographic trends. The CFC concentrations module is a two box model based on Wigley (1988).

With this simple CFC two box model the effect of the Montreal Ozone Protocol on future concentrations of CFCs are calculated and evaluated.

6.2 CFCs Emissions Module

Historical production figures of CFC-11 and CFC-12 are gathered by the Chemical Manufacturers Association (CMA, 1987). In Figures 6.1 and 6.2 historical production numbers of CFC-11 (from 1931) and CFC-12 (from 1939) are given. These numbers contain many imponderables (Rotmans, 1986). The model disaggregates production of CFCs into four major categories of end use, partly based on Gamlen et al.(1986): aerosols and open cell foams, closed cell foams, hermetically and non-hermetically sealed refrigerators. For these four categories manufacturing losses, direct emissions and delayed emissions are calculated.

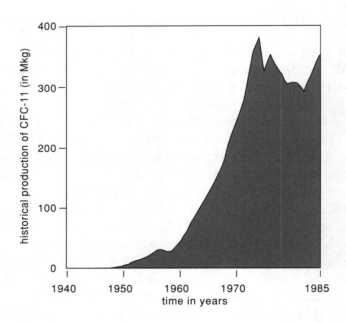

Figure 6.1: Historical production of CFC-11.

6.2. CFCS EMISSIONS MODULE

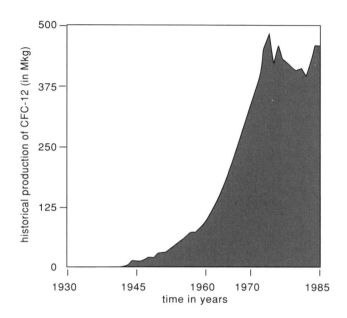

Figure 6.2: Historical production of CFC-12.

The subdivision of chlorofluorocarbons into end use categories is described by the following equation:

$$PRD(t) = S_{aroc}(t) + S_{cc}(t) + S_{hsrfr}(t) + S_{nhsrfr}(t) \qquad (6.1)$$

with:
$PRD(t)$ = production of CFC at time t (in MkgCFC)
$S_{aroc}(t)$ = sales of aerosols and open cell foams at time t (in MkgCFC)
$S_{cc}(t)$ = sales of closed cell foams at time t (in MkgCFC)
$S_{hsrfr}(t)$ = sales of hermetically sealed refrigerators at time t (in MkgCFC)
$S_{nhsrfr}(t)$ = sales of non-hermetically sealed refrigerators at time t (in MkgCFC)

Equation (6.1) only holds for countries which report to the CMA. The uncertainties in the production data from the CMA reporting companies are probably not more than ± 0.5% (Gamlen et al., 1986). For other countries (non CMA-participating countries which produce CFCs, like the USSR, China, India and Argentina) the estimated range of uncertainties is much larger. This causes an uncertainty in cumulative release figures for all CFC producing countries of about 4% for CFC-11 and about 2.8% for CFC-12 according to Gamlen et al. (1986). For the non-reporting countries only a distinction is made between aerosols and non-aerosols. The historical production of the non-reporting countries is known for the period 1965–1975, while for the period 1975–1986 an annual growth of 8% is supposed. Future production for the period 1986–2100 is presumed to follow the scenarios for the CMA reporting countries, which are described below.

Four CFC use scenarios have been distinguished, see Rotmans et al. (1990a). In the *unrestricted trends* (A) scenario it is assumed that the Montreal protocol (United Nations Environment Program, 1987) is not fully implemented: production and use will be increasing, albeit at a moderate rate, and will stabilize in the first half of the 21st century.

In the *reduced trends* (B) scenario the Montreal protocol is supposed to be implemented by linearly decreasing the production towards the limits set for 1993 (80%) and 1998 (50%), followed by stabilization after 1999. The allowed growth of planned production is not taken into account because of a lack of data. The objectives are reached by turning to substitutes, limiting losses and gradually increasing recycling.

In the *changed trends* (C) scenario the basic protocol is upgraded to a limit of 15% of the 1986 production and consumption figures for 1998, after which annual production and consumption are assumed to stabilize. No additional production capacity is built, since this seems to be useless under these circumstances. The efforts towards the development of 'safe' substances, recycling and loss reduction are increased.

In the *forced trends* (D) scenario a total phase-out is assumed in 2050 in addition to the 85% reduction in the changed trends scenario (C). A possibly remaining use for purposes considered to be essential is neglected. The CFC emissions are split up into prompt emissions and delayed emissions. The prompt CFC emission is composed of fugitive emission, aerosol and open cell foam emission, closed cell foam emission, refrigerator emission and non-CMA reported emissions. Each end use category has a different lifetime, represented in the model by specific fractions. These constants are empirically based on the lifetime of the different products, which is less than 1 year for aerosols and open cell foams, about 4 years for closed cell foams, about 4 years for non-hermetically sealed refrigerators and about 12 years

6.2. CFCS EMISSIONS MODULE

for hermetically sealed refrigerators (Gamlen et al., 1986).

Not all the CFC-11 and CFC-12 manufactured is sold. Some is lost immediately during the production process, with no delay between production and release. These fugitive emissions are estimated to be 2% of CFC-11 production and 3.3% of CFC-12 production (CMA, 1983). The following equation is used to estimate the prompt emissions of CFCs:

$$\begin{aligned} EMCFC(t) =\ & EM_{fug}(t) + PRD_{aroc}(t) \\ & + ALFCCF * PRD_{ccf}(t) \\ & + ALFRFN * PRD_{nhsrfr}(t) + ALFRFH * PRD_{hsrfr}(t) \\ & + PRD_{aronc}(t) + ALFOC * PRD_{naronc}(t) \end{aligned} \tag{6.2}$$

with:

$EMCFC(t)$	=	emission of CFCs at time t (in Mkg)
$EM_{fug}(t)$	=	fugitive emission of CFCs
$PRD_{aroc}(t)$	=	production of CFCs by aerosols and open cell foams at time t
$ALFCCF$	=	fraction of closed cell foams production released to the atmosphere (equivalent to the inverse average life time of closed cell foams, 1/4)
$PRD_{ccf}(t)$	=	production of CFCs by closed cell foams at time t
$ALFRFN$	=	fraction of non-hermetically sealed refrigerators released to the atmosphere (inverse average lifetime of non-hermetically sealed refrigerators, is 1/4)
$PRD_{nhsrfr}(t)$	=	production of CFCs by non-hermetically sealed refrigerators time t
$ALFRFH$	=	fraction of hermetically sealed refrigerators released to the atmosphere (only for CFC-12, is 1/12)
$PRD_{hsrfr}(t)$	=	production of CFCs by hermetically sealed refrigerators at time t
$PRD_{aronc}(t)$	=	not CMA reported production of CFCs by aerosols at time t
$ALFOC$	=	average fraction of not CMA reported production of CFCs by non-aerosol applications released to the atmosphere (is 1/4)
$PRD_{naronc}(t)$	=	not CMA reported production of CFCs by non-aerosol applications at time t

Besides prompt emissions of CFCs the model also contains delayed emissions, accounting for the fraction of production of CFCs that does not immediately escape into the atmosphere. These delayed emissions are modelled by a delayed-release box according to Wigley (1988):

$$\begin{aligned}
DLEMCF(t) = \ & DLEMCF(t-1) + \int_{t-1}^{t} [(1-ALFRFN) * PRD_{nhsfr}(\tau) \\
& + (1 - ALFRFH) * PRD_{hsrfr}(\tau) \\
& + (1 - ALFCCF) * PRD_{ccf}(\tau) \\
& + (1 - ALFOC) * PRD_{naronc}(\tau) \\
& - CFCBET * DLEMCF(\tau)] d(\tau) \quad (6.3)
\end{aligned}$$

with:
$DLEMCF(t)$ = delayed emission of CFCs (in Mkg)
$DLEMCF(0)$ = delayed emission at time $t = 0$, in 1931, and is 0
$CFCBET$ = annual fraction that leaks from the delayed release box into the atmosphere box; calibrated on historical data, is 0.08 for CFC-11 and 0.15 for CFC-12)

6.3 CFCs Concentrations Module

The concentration model used is a simple representation of the complicated atmospheric chemistry and comprises a second box which is directly linked to the delayed emissions box. The removal process of CFCs is assumed to be inversely proportional to the atmospheric lifetime of CFCs, the latter being constant. Estimates for lifetimes of CFCs are given in Bolle et al. (1986), Cunnold et al. (1983a and 1983b), Ramanathan et al. (1985) and Wigley (1988): they are 75 years for CFC-11 with a range of uncertainty 50–107 years, and 125 years for CFC-12 with a range of uncertainty 65–400 years. The tropospheric concentration of CFCs is determined by the initial concentration of CFCs, their emissions and their removal. CFCs are assumed to be removed from the atmosphere due to stratospheric loss only and in proportion to their concentration. Then the fraction of CFCs remaining at time t is reflected by the negative exponential function $e^{-t/LFTCFC}$, where LFTCFC is the atmospheric lifetime of CFCs.

The global tropospheric concentration is then calculated as follows:

$$PCFC(t) = PCFC(t-1) + \int_{t-1}^{t} [(CVFCFC * EMCFC(\tau)$$
$$- PCFC(\tau)/LFTCFC$$
$$+ CVFCFC * CFCBET * DLEMCF(\tau)]d(\tau) \quad (6.4)$$

with:
$PCFC(t)$ = tropospheric concentration of CFCs at time t (in ppt)
$PCFC(0)$ = initial tropospheric concentration of CFCs, is 0
$CVFCFC$ = conversion factor of CFCs (in ppt/Mkg) is 0.047 ppt/Mkg CFC-11 and is 0.050 ppt/Mkg CFC-12, according to Chamberlain et al. (1982)
$EMCFC(t)$ = emission of CFCs at time t (in Mkg/yr)
$LFTCFC$ = atmospheric lifetime of CFCs (75 year for CFC-11 and 125 year for CFC-12, assumed to be constant
$DLEMCF(t)$ = delayed emission of CFCs (in Mkg/yr)
$CFCBET$ = annual fraction that leaks from the delayed release box into the atmosphere box; calibrated on historical data, is 0.08 for CFC-11 and 0.15 for CFC-12)

6.4 Results

Figures 6.3 and 6.4 shows these production scenarios for CFC-11 and CFC-12, respectively. These production scenarios are somewhat higher than those given in Rotmans et al. (1990a). Additionally, production scenarios for another controlled CFC is represented, CFC-113. CFC-113 is widely used as a solvent in the electronics industry.

Historical production numbers of CFC-113 are derived from GEMS (1989), and for the period 1986–1990 a growth factor of 7.6 percent annually is used (Hammitt et al., 1986). Figure 6.5 represents the CFC-113 production scenarios.

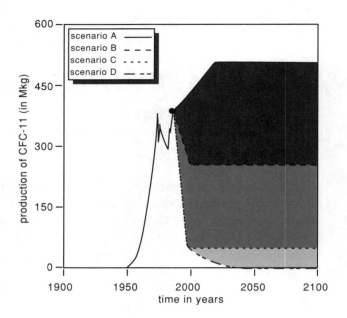

Figure 6.3: Production of CFC-11.

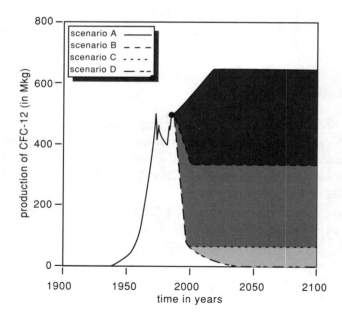

Figure 6.4: Production of CFC-12.

6.4. RESULTS

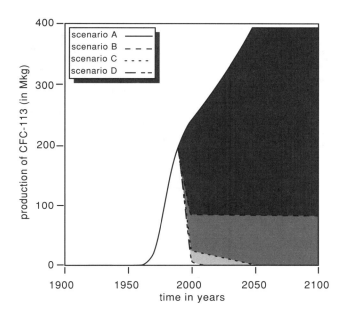

Figure 6.5: Production of CFC-113.

Figures 6.6 and 6.7 give the concentrations for CFC-11 and CFC-12, respectively. Although the influence is undeniable, the UNEP Ozone Protocol cannot prevent CFC-11 and CFC-12 concentrations from increasing over the next decades.

Only by upgrading the Montreal Protocol (scenario C, the changed trends) the concentrations of CFC-11 and CFC-12 can be maintained at the present level. Figure 6.8 gives the concentration of CFC-113. In spite of control by the Montreal Protocol scenario A yields a concentration value in 2100 of more than 20 times the 1985 concentration value. Only scenarios C and D lead to concentrations in 2100 that are comparable with that of 1985.

These results may therefore even be considered to be optimistic. Due to the long atmospheric lifetime of CFCs still further reduction of emissions (scenario D, the forced trends) is needed to lower CFC concentrations in the second half of the next century.

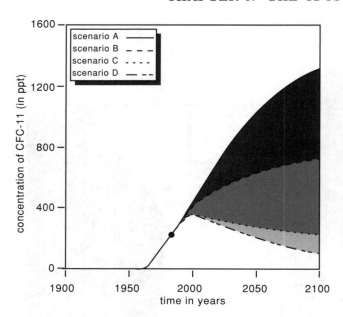

Figure 6.6: Concentration of CFC-11.

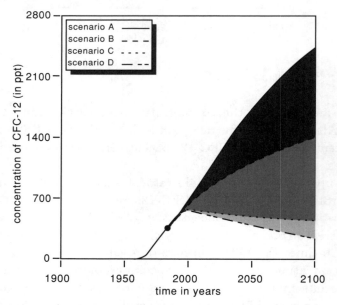

Figure 6.7: Concentration of CFC-12.

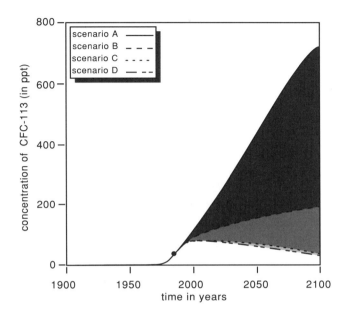

Figure 6.8: Concentration of CFC-113.

6.5 Conclusions

In light of simulation results with IMAGE it can be concluded that the effect of the implementation of the Montreal Ozone Protocol is important to stabilize the relative role of CFCs in the greenhouse effect and is a significant step forward. However under the present Protocol conditions CFC concentrations can still increase considerably. To eliminate the role of CFCs in the greenhouse effect it is necessary to strongly upgrade the ozone agreement of Montreal.

Chapter 7

The Climate Module

7.1 Introduction

The very complex climate system involves transfers of energy between the atmosphere and land surface, oceans, biomass and cryosphere. Our present understanding of this climate system is still not satisfactory. Climate models try to consider the primary processes, radiation, dynamics and surface processes, and their interactions within the climate system (Shine and Henderson-Sellers, 1983, Dickinson, 1986). Climate modelling has developed rapidly in recent decades. A broad range of such climate models have been developed, varying from simple zero-dimensional to elaborate three dimensional models (Hasselman, 1988, and Washington and Parkinson, 1986). Generally four climate model types may be are distinguished:

1. Energy Balance Models (EBM): EBMs predict the change in temperature at the earth's surface resulting from a change in heating based on the requirement that the net flux of energy does not change. The largest limitation of EBMs is that they do not have a physically based model of the atmosphere (Schlesinger, 1985);

2. Radiative Convective Models (RCM): RCMs compute the vertical (usually globally averaged) temperature structure of the atmosphere from the balance between radiative heating or cooling and the vertical heat flux; Because RCMs are only "average" models of the global climate system, the quantitative model results of RCMs strongly depend upon the assumptions (cloud feedback, surface albedo) and therefore are not very reliable (Shine and Henderson-Sellers, 1983, Schlesinger, 1986);

3. Two-dimensional statistical dynamical (SD) models mostly combines the latitudinal dimension of the energy balance models with the vertical dimension of the radiative-convective models. A set of statistics is used to represent the wind speeds and directions. Although they provide insight into the complex climate system, their use in climate prediction is rather limited because of the lack of zonal resolution (Henderson-Sellers and Verstraete, 1987);

4. General Circulation Models (GCM): GCMs calculate the changes in atmospheric dynamics, without the restriction of explicit spatial averaging. Because the full three-dimensional nature of the atmosphere is resolved, trying to represent all important physical processes, GCMs are the most sophisticated models currently available. These models have a horizontal resolution of several hundred kilometers. The two major problems of GCMs are the limits upon computer time and adequate data sets (Schlesinger, 1985, Shine and Henderson-Sellers, 1983). A further distinction is often drawn between oceanic general circulation models (OGCMs) and atmospheric general circulation models (AGCMs). Coupled atmosphere-ocean general circulation models (A/O GCMs) have a coarse resolution and are very costly to run. In addition only a very small number of simulations have been performed with these models (Harvey, 1989a).

In Figure 7.1 these different types of climate models are represented by the climate modelling pyramid. The higher its position up the pyramid the more complex a climate model is and the greater are the interactions between the primary processes radiation, dynamics and surface processes.

In spite of their deficiencies simple climate models are very useful. On the one hand they enhance insight in the intricate climate processes and are a tool for interpreting more detailed models, while on the other their flexibility compared to more elaborate models, allows wide variation in parameter values (Dickinson, 1986). IMAGE uses a one-dimensional energy balance model, based on Wigley and Schlesinger (1985), which will be discussed in this chapter.

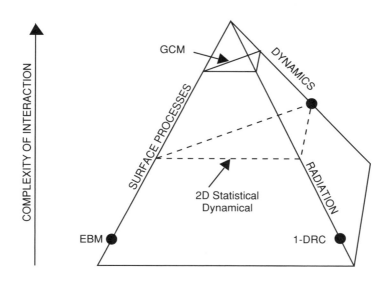

Figure 7.1: The climate modelling pyramid (from Shine and Henderson-Sellers, 1983).

7.2 Model Description

7.2.1 Equilibrium Response

The simplest useful climate model which can be written is (Dickinson, 1986):

$$C_m * \frac{d\Delta T}{dt} = \Delta Q - \lambda * \Delta T \tag{7.1}$$

with:
C_m = effective heat capacity of the earth (in $W.y/m^2.°C$)
ΔT = change in global surface temperature from some climatological value (in $°C$)
ΔQ = perturbation in the energy flux to the surface (in W/m^2)

λ = climate sensitivity parameter LAMBDA; $\lambda = \frac{\Delta F}{\Delta T} - \frac{\Delta S}{\Delta T}$, where F and S respectively denote the global mean emitted infrared and net downward solar fluxes at the top of the atmosphere. Thus λ represents a change in climate due to a given radiative forcing ΔQ. λ itself depends on the integrated effect of a number of feedback processes involving clouds, surface albedo, and the atmospheric vertical temperature and humidity structure (in $W/m^2.°C$).

In equilibrium, supposing an instantaneous response of global temperature to external radiative forcing, the first term is zero, yielding the following equation, which is modelled in IMAGE:

$$\Delta T_e = \Delta Q/l \tag{7.2}$$

with:
ΔT_e = the change of the global mean surface temperature (in $°C$) in the equilibrium phase

This equilibrium temperature is induced by the radiative perturbation (forcing) of the surface-troposphere system, caused by the increase in trace gas concentrations. In order to calculate this equilibrium temperature increase the total radiative forcing must be determined. Knowing the different trace gas concentrations for CO_2, CH_4, N_2O, CFC-11 and CFC-12, the radiative forcing can be calculated. To estimate the total change in radiative forcing at the tropopause (ΔQ) induced by the trace gases, the total forcing is modelled as follows (Wigley, 1985, 1987):

$$\Delta Q = \Delta Q_{CO_2} + \Delta Q_{CH_4} + \Delta Q_{N_2O} + \Delta Q_{CFC-11} + \Delta Q_{CFC-12} \tag{7.3}$$

with:
ΔQ = total change in radiative forcing (in W/m^2)
$\Delta Q_{CO_2}, \Delta Q_{CH_4}, \Delta Q_{N_2O}$
$\Delta Q_{CFC-11}, \Delta Q_{CFC-12}$ = change in radiative forcing by CO_2, CH_4, N_2O, CFC-11 and CFC-12 (in W/m^2)

Following Wigley (1987) the relationship between change in radiative forcing and change in concentration is assumed to be linear at low concentrations, square root at intermediate values and logarithmic at high concentrations. Approximately this yields a logarithmic relation for CO_2, a square root relation for CH_4 and N_2O, and a linear relation for the CFCs. According to Ramanathan et al. (1979) the following approximate relation holds for CO_2:

7.2. MODEL DESCRIPTION

$$\Delta Q_{CO_2} = (\Delta Q_{2xCO_2}/Ln(2)) * Ln(pCO_2/pCO_2in) \quad (7.4)$$

with:

ΔQ_{2xCO_2} = radiative forcing for a doubled CO_2 concentration; is 4.3 W/m^2 according to Ramanathan et al., 1985, and Hansen et al., 1988
pCO_2 = atmospheric CO_2 concentration (in *ppm*)
pCO_2in = pre-industrial atmospheric CO_2 concentration, in 1900 (in *ppm*)

Augustsson and Ramanathan (1977) state that this relation is valid only for CO_2 concentrations within the interval between the present and four times the present CO_2 concentration. For CH_4, the change in radiative forcing is estimated to be proportional to the square root of the concentrations of CH_4 and N_2O, whereas the proportionality constant 0.0398 (in $W/(m^2.ppb)$) is based on model results of Kiehl and Dickinson (1987):

$$\Delta Q_{CH_4} = 0.0398 * (\sqrt{pCH_4} - \sqrt{pCH_4in}) \quad (7.5)$$

with:

pCH_4 = atmospheric CH_4 concentration (in *ppb*)
pCH_4in = pre-industrial atmospheric CH_4 concentration, in 1900 (in *ppb*)

This relation is fitted with the Kiehl-Dickinson model over the range 700 *ppb* to 3100 *ppb*. In IMAGE this relation is also used for CH_4 concentrations falling outside this range. However, this seems to be permitted in light of the fact that the radiative forcing functions are probably accurate to about 10% (Wigley, 1987). For N_2O a square root relationship is supposed as well, following Ramanathan et al. (1985):

$$\Delta Q_{N_2O} = 0.105 * (\sqrt{pN_2O} - \sqrt{pN_2Oin}) \quad (7.6)$$

with:

pN_2O = atmospheric N_2O concentration (in *ppm*)
pN_2Oin = pre-industrial atmospheric N_2O concentration, in 1900 (in *ppm*)

For CFC-11 and CFC-12 a linear relation is used, with linear proportionality constants of Ramanathan et al. (1985):

$$\Delta Q_{CFC-11} = 0.27 * pCFC-11 \tag{7.7}$$

with:
$pCFC-11$ = atmospheric CFC-11 concentration (in *ppm*)

$$\Delta Q_{CFC-12} = 0.31 * pCFC-12 \tag{7.8}$$

with:
$pCFC-12$ = atmospheric CFC-12 concentration (in *ppm*)

Following Wigley (1985) and Tricot and Berger (1987) an equivalent CO_2 concentration can be defined, including the combined radiative forcing of all these trace gases (CO_2, CH_4, N_2O, CFC-11 and CFC-12). This CO_2 equivalent concentration expresses the total effect of all trace gases compared to the effect of CO_2 alone. Using Equation (7.4) the equivalent CO_2 concentration can be defined as:

$$\Delta Q = (\Delta Q_{2xCO_2}/Ln(2)) * Ln(pCO_2eq/pCO_2) \tag{7.9}$$

with:
pCO_2eq = CO_2 equivalent concentration (in *ppm*)

Substituting Equation (7.3) in this relation yields the following expression:

$$pCO_2eq = pCO_2 * e^{[Ln(2)/\Delta Q_{2xCO_2}*(\Delta Q_{CH_4}+\Delta Q_{N_2O}+\Delta Q_{CFC-11}\Delta Q_{CFC-12})]}$$

$$\tag{7.10}$$

with:
pCO_2eq	= CO_2 equivalent concentration (in *ppm*)
pCO_2	= atmospheric CO_2 concentration (in *ppm*)
ΔQ_{2xCO_2}	= radiative forcing for a doubled CO_2 concentration (in W/m^2)
$\Delta Q_{CH_4} + \Delta Q_{N_2O} +$ $\Delta Q_{CFC-11} + \Delta Q_{CFC-12}$	= radiative forcing, resulting from changes in CH_4, N_2O, CFC-11, and CFC-12 concentration (in W/m^2)

Equation (7.10) can be expressed in another way:

7.2. MODEL DESCRIPTION

$$pCO_2eq = pCO_2in * e^{[(Ln(2)/\Delta Q_{2xCO_2} * \Delta Q]} \tag{7.11}$$

with:
pCO_2in = pre-industrial CO_2 concentration (in ppm)
ΔQ = total radiative forcing, caused by changes in concentrations of all trace gases, as defined in Equation (7.3) (in W/m^2)

The CO_2 equivalent concentration, as expressed in (7.10) and (7.11), defines a CO_2 concentration of which the radiative forcing is as large as the combined effect of the trace gases CO_2, CH_4, N_2O, CFC-11 and CFC-12. In the framework of the Intergovernmental Panel on Climate Change (IPCC) the CO_2 equivalent concentration functions as a target indicator, as shown in Chapter 9.

7.2.2 Transient Response

The reconstructed global mean temperature increase from 1900 to 1985 indicates a warming effect of $0.5 - -0.7°C$ (Schlesinger, 1986, Wigley, 1987, and Hansen et al., 1988). However, climate model calculations show a global mean equilibrium temperature increase which is about twice as much. A main contributor to this difference is supposed to be the phenomenon that the actual transient (delayed) response of the climate systems lags the equilibrium response because of the thermal inertia of the ocean (Schneider and Thompson, 1981). Also other factors might offset the greenhouse effect on global mean temperature, such as the internal climate variability, variations of solar radiation, aerosols and albedo changes (IPCC, 1990). The transient temperature effect can be calculated with an energy balance box-diffusion model, based on Wigley and Schlesinger (1985). This climate model includes a land box, an ocean box and atmosphere boxes over land and ocean. The basic equation of the energy balance model is the following:

$$C_m * \frac{d\Delta T}{dt} = \Delta Q - \lambda * \Delta T - \Delta F \tag{7.12}$$

with:
C_m = bulk heat capacity of the ocean mixed layer (in $W.y/m^2.°C$)
ΔT = change in temperature of the ocean mixed layer (in $°C$)
ΔQ = perturbation in the energy flux to the surface (in W/m^2)
λ = climate sensitivity parameter, see Equation (7.1)
ΔF = change in heat flux at the bottom of the mixed layer (in W/m^2)

The ΔF can be calculated by the following equation:

$$\Delta F = -C_m * DIFF/D * [\frac{\delta \Delta T_o}{\delta z}]_{z=0} \qquad (7.13)$$

with:
- ΔF = change in heat flux at the bottom of the mixed layer (in W/m^2)
- C_m = effective heat capacity of the mixed ocean layer; (in $W.y/m^2.°C$) is estimated as (7.4) $W.y/m^2.°C$, based on Wigley and Schlesinger (1985)
- $DIFF$ = diffusion coefficient; is 4000 m^2/yr, based on Goudriaan and Ketner (1984); varies in the literature from about 3700 or 6000 m^2/yr, according to Hoffman (1984), see Chapter 2 (in m^2/yr)
- D = depth of the mixed layer of the ocean (in m)
- ΔT_o = change in temperature of the deep ocean (in $°C$)
- z = depth below the bottom of the mixed layer (deep ocean) (in m)

The ΔT_0 is determined by the following diffusion equation:

$$\frac{\delta \Delta T_0}{\delta t} = DIFF * \frac{\delta^2 \Delta T_0}{\delta z^2} \qquad (7.14)$$

with the boundary conditions $\Delta T_0 = \Delta T$ when $z = 0$, and $\Delta T_0 = 0$ when z goes to infinity. While Wigley and Schlesinger (1985) give an approximate analytical solution ΔT_0, in the climate model of IMAGE the delayed temperature effect is calculated in two ways. The first solution is the numerical approach, where the partial differential equation is solved by writing it in a discrete way. The second solution involves an analytical approach, following Wigley and Schlesinger (1985). Only the numerical approach will be discussed here. Equation (7.13) is therefore written as:

$$\Delta F = -C_m * DIFF/D * (\frac{\Delta T_0(1) - \Delta T}{\Delta z}) \qquad (7.15)$$

with:
- ΔF = change in heat flux at the bottom of the mixed layer (in W/m^2)
- $\Delta T_0(1)$ = change in temperature of first deep ocean layer ($°C$)
- ΔT = change in temperature of the ocean mixed layer (in $°C$)
- Δz = thickness of a certain deep ocean layer (in m)

while Equation (7.14) is written as:

$$\frac{d\Delta T_0(i)}{dt} = DIFF * (\frac{\Delta T_0(i-1) - 2 * \Delta T_0(i) + \Delta T_0(i+1)}{\Delta z^2})$$
$$i = 1, \ldots, n-1 \qquad (7.16)$$

7.2. MODEL DESCRIPTION

with:
$\Delta T_0(i)$ = change in temperature in deep ocean layer i, $i = 1, \ldots, n-1$ with $\Delta T_0(n) = 0$

In matrix-notation:

$$\begin{bmatrix} \Delta T_0(1) \\ \vdots \\ \vdots \\ \vdots \\ \vdots \\ \vdots \\ \Delta T_0(n) \end{bmatrix} = DIFF/\Delta z^2 * \begin{bmatrix} 1 & -2 & 1 & 0 & \ldots & 0 \\ 0 & 1 & -2 & 1 & \ldots & 0 \\ 0 & \ldots & \ldots & \ldots & \ldots & \ldots \\ \ldots & \ldots & \ldots & \ldots & \ldots & \ldots \\ \ldots & \ldots & \ldots & \ldots & \ldots & \ldots \\ \ldots & \ldots & \ldots & 1 & -2 & 1 \\ 0 & 0 & \ldots & \ldots & \ldots & 0 \end{bmatrix} \begin{bmatrix} \Delta T_0(0) \\ \Delta T_0(1) \\ \Delta T_0(2) \\ \ldots \\ \ldots \\ \Delta T_0(n-1) \\ 0 \end{bmatrix}$$

(7.17)

Then the temperature change of the atmosphere over land can be calculated. Assuming that the temperature change of the atmosphere over the ocean equals the temperature change in the mixed ocean layer, the surface-air temperature change can be expressed as (Wigley and Schlesinger, 1985)

$$\Delta T_{sa} = \frac{f \lambda \Delta T_{eq} + k \Delta T}{f \lambda + k} \quad (7.18)$$

with:
ΔT_{sa} = change in surface-air temperature ($^\circ C$), which in the following is called transient air temperature
ΔT_{eq} = change in equilibrium temperature ($^\circ C$)
ΔT = change in temperature of the ocean mixed layer ($^\circ C$)
f = fraction of the global covered by land
λ = climate sensitivity parameter, see Equation (7.1) ($W/m^2~^\circ C$)
k = coefficient that represents the heat transfer between land and ocean.

This model fully parameterizes the exchange of heat between the different boxes.

Concerning the ocean box, a more elaborate version with diffusion and upwelling, has been implemented, based on Michael et al. (1981). Simulation results with this model cannot yet be presented.

In order to obtain a stable solution, the following stability requirement must hold:
$$(DIFF * \Delta t)/\Delta z^2 \leq 0.5 \qquad (7.19)$$

with:
$DIFF$ = diffusion coefficient; is 4000 m^2/yr, as stated above (in m^2/yr)
Δt = simulation time step (in yr)
Δz = thickness of a deep ocean layer (in m)

Taking a Δz value of 100 meter, and DIFF = 4000, then Δt must be less than 1.25 year. This is in line with the general simulation time step of 0.5 year for the whole simulation program (see Chapter 2). Presuming a simulation time step of 0.5 year, the minimum thickness value appears to be about 63 meter. Otherwise a Δz value of about 15 meter, as taken by Hoffman et al. (1983), requires a maximum simulation time step of about 0.03 year (or 10 days).

A number of simulation experiments have been made with Δz varying from 15 to 200 meters. It appeared that, by choosing a Δz value of 25 meters, the numerical solution approximates the analytical solution. The simulation time step must then at any rate be less then 0.078 year, but simulation experiments showed a time step of 1/20 year to be necessary and sufficient to get a stable solution, and to approximate the analytical solution.

The disadvantage of such a small time step is the huge amount of computer time it requires. Following Spelman and Manabe (1984), who suggest that, on a time scale of several decades, only the deep layer down to 1000 meters will be warmed up, resulting in a mixed ocean layer of 75 meters and a deep layer of 925 meters, the deep ocean layer is divided into 37 layers of 25 meters each.

The thermal diffusivity is assumed to be constant (although it is dependent on depth and temperature), and is used as a substitute for complex circulation processes that transport heat. This may lead to an overestimation of downward heat transport (Hoffman et al., 1983). To investigate the role of the diffusion coefficient a sensitivity analysis has been performed with a range of diffusion coefficients used, 3700 to 6000 m^2/yr. Additionally extreme values, falling outside this used range, are even possible, due to the temperature dependence of the thermal diffusivity coefficient. In this way a temperature increase may induce extreme values of the thermal diffusion coefficient. Various simulation experiments have been performed with different values of the thermal diffusivity, which are presented in the results section.

7.2. MODEL DESCRIPTION

The ratio $\Delta T/\Delta T_{eq}$ (with T the delayed or transient temperature effect, and T_{eq} the equilibrium temperature effect) is a measure of the degree of equilibrium. In equilibrium phase $\Delta T/\Delta T_{eq}$ is 1, while deviation from 1 indicates a disequilibrium phase (in the results section the ratio $\Delta T/\Delta T_{eq}$ is plotted for different scenarios). The ocean surface response time (time to reach $1-e^{-1}$ of equilibrium response) is strongly dependent on the climate feedback factor λ (Hansen et al., 1985).

7.2.3 Climate Feedbacks

One of the largest gaps in our knowledge of the greenhouse problem concerns the climate feedback mechanism, often specified by the climate feedback factor or, equivalently, by the equilibrium CO_2 doubling temperature change. A doubling of the CO_2 concentration, taking no account of feedbacks, would cause a global mean surface temperature increase of about $1.2°C$, according to various model calculations (Schlesinger, 1986, Schuurmans et al., 1982). However, taking feedback mechanisms into account, a doubling CO_2 temperature increase lies in the range $1.5 - -4.5°C$. This wide range is caused by the many uncertainties about feedback mechanisms in the climate system, often pointed to as the climate sensitivity. Kellog (1983) distinguishes five climate feedback loops, of which three are considered as positive and two as negative.

Following Lashof (1989), feedbacks can be subdivided into geophysical and biogeochemical feedbacks. The most important geophysical feedbacks are the water vapor feedback factor, the cloud feedback, and the ice and snow feedback. In contrast to these feedbacks, which are involved in current climate models, the biogeochemical feedbacks have not yet been taken up in these models.

Major biogeochemical feedbacks are the release of methane hydrates, changes in ocean chemistry, and changes in the vegetation albedo. Lashof (1989) tries to quantify the gain from the biogeochemical feedbacks, which is about 0.05–0.29, compared with 0.17–0.77 for the geophysical feedbacks. Table 7.1 gives an overview of the various feedbacks:

One has to be very careful with adding all these contributions in Table 7.1, because of the different time scales of the separate feedbacks. Apart from this, the given total gain is only indicative, because the sum of the separate feedback gains does not equal to the total gain. Clearly, more research is urgently needed to reduce the wide range of climate sensitivities.

GEOPHYSICAL FEEDBACK	GAIN	
Water Vapor	0.39	(0.28–0.52)
Ice and Snow	0.12	(0.03–0.21)
Clouds	0.09	(−0.12–0.29)
Sub-Total	0.64	(0.17–0.77)
BIOGEOCHEMICAL FEEDBACK		
Methane Hydrates	0.10	(0.01–0.20)
Tropospheric Chemistry	−0.04	−(0.01–0.06)
Ocean Chemistry	0.008	
Ocean Eddy-Diffusion	0.02	
Ocean Biology & Circulation	0.06	(0.00–0.10)
Vegetation Albedo	0.05	(0.00–0.09)
Vegetation Respiration	0.01	(0.00–0.03)
CO_2 Fertilization	−0.02	−(0.01–0.04)
Methane from Wetlands	0.01	(0.003–0.015)
Methane from Rice	0.006	(0.00–0.01)
Electricity Demand	0.001	(0.00–0.004)
Sub-Total	0.16	(0.05–0.29)
Total	0.80	(0.32–0.98)

Table 7.1: Estimated gain from geophysical and biogeochemical feedbacks, from Lashof (1989).

7.3 Results

As mentioned before the feedback parameter remains a crucial source of uncertainty, resulting in a wide range of $1.5 - -4.5°C$ for the global CO_2 doubling equilibrium temperature rise (ΔT_{2xCO_2}). Therefore simulation results will be presented for different values of ΔT_{2xCO_2}, viz. $\Delta T_{2xCO_2} = 2.0, 3.0$ and $4.0°C$.

The results of combined calculations, as shown in Figure 7.2, are quite surprising. Choosing a value of $3.0°C$ for ΔT_{2xCO_2}, even with the most restrictive emissions scenarios a 2.0 $°C$ increase of the equilibrium temperature will occur before 2020. Actually for 1985 a global equilibrium temperature increase of 1.4 $°C$ has been simulated, which is in line with other models (Dickinson, 1986, Wigley, 1987). Double the amount of CO_2 in scenario A results in an increase of 3.0 $°C$ in about 2050, while the total effect then of all trace gases is a 5.0 $°C$ increase.

7.3. RESULTS					135

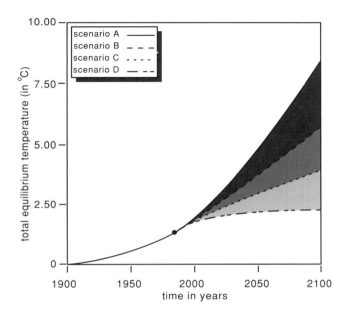

Figure 7.2: Global mean equilibrium temperature increase for different scenarios, with $\Delta T_{2xCO_2} = 3.0°C$.

Figure 7.3 shows the resulting combined equilibrium temperature response, but then with $\Delta T_{2xCO_2} = 2.25\ °C$. This figure yields a global equilibrium temperature increase of 4.0 $°C$ in case of doubling the CO_2 concentration.

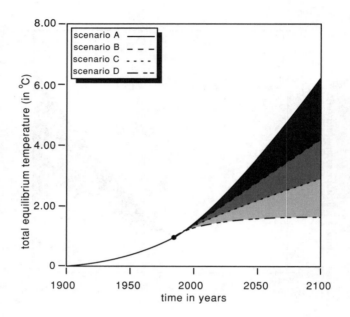

Figure 7.3: Global mean equilibrium temperature increase for different scenarios, with $\Delta T_{2xCO_2} = 2.25°C$.

Figure 7.4 shows the global equilibrium temperature increase for the unrestricted trends scenario, for ΔT_{2xCO_2} values of 2.0, 3.0 and 4.0 $°C$ respectively. Choosing a climate sensitivity (ΔT_{2xCO_2}) of 2.0 or 4.0 $°C$ gives a difference in total equilibrium temperature rise of more than 4.0 $°C$ in the year 2100, again stressing the importance of the varied climate sensitivity.

7.3. RESULTS

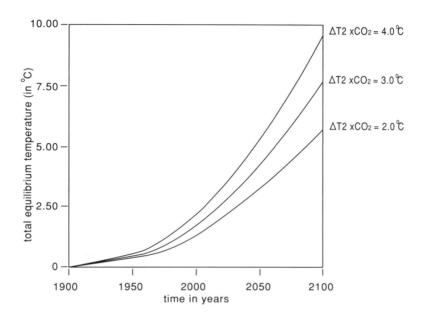

Figure 7.4: Global mean equilibrium temperature increase for the unrestricted trends scenario, for CO_2 doubling temperature increase values of 2.0, 3.0 and 4.0 $°C$, respectively.

The global transient temperature response is represented in Figure 7.5. For 1985 a transient response is simulated of 0.6 $°C$, which corresponds with measured global mean temperature rises, varying from 0.5 to 0.7 $°C$ (Schlesinger, 1986, Wigley, 1987, and Hansen et al. 1988). For 2100 the transient temperature response varies from about 5 $°C$ for the unrestricted trends scenario (A) to about 1.5 $°C$ for the forced trends scenario (D).

In Figure 7.6 the equilibrium and transient air temperature response over the period 1900–1985 are both depicted. The total equilibrium response in 1985 is about 1.4 $°C$, while the transient response in 1985 is about 0.6 $°C$. From this figure it follows that more than half of the expected global warming effect has not yet been realized, given a sensitivity (or ΔT_{2xCO_2}) of 3.0 $°C$.

In Figure 7.7 both the equilibrium and transient response are depicted, for a moderate scenario, a changed trends scenario, leading to an equilibrium temperature difference of about 1.3 $°C$.

Figure 7.8 gives both the surface air transient and ocean mixed layer temperature response. In particular, until 2050 they hardly differ from each other, and only thereafter does the difference become significant; this is caused by the parametrization chosen. Other realistic parameterizations will lead to various differences between the surface air and ocean response, although these differences will not alter dramatically.

Figure 7.9 gives the relative contribution of the individual trace gases to equilibrium temperature rise, which varies over the scenario sets and is time dependent. The contribution is simulated for two scenario sets. A and D, while three years are chosen, 1985, 2050 and 2100. The results presented are not cumulative, but are calculations for one year. This implies that the contribution of a certain trace gas in an arbitrary year indicates the difference in temperature increase between the previous year and the chosen arbitrary year.

In both scenarios CO_2 is the dominant greenhouse gas. The present contribution of CO_2 is found to be about 60%. The contributions of both CO_2 and the other gases for 1985 are in line with the figure presented by Hansen et al. (1988). The contribution of the CFCs to the total temperature effect is remarkable. A present 22% contribution is found to decrease to about 11% in 2050 to 8% in 2100 for scenario A to 0% in scenario D. This means that the role of CFCs, in spite of the Montreal Protocol, is not played out yet.

Table 7.10 gives the CO_2 equivalent concentration, as defined in (7.10) and (7.11), for the different scenarios. The CO_2 equivalent doubling value, which is 570 *ppm* in IMAGE (starting from a value of 285 *ppm*), is reached in 2030 for scenario A, the unrestricted trends, in 2040 for scenario B, in 2070 for scenario C and will not be reached for scenario D.

7.3. RESULTS

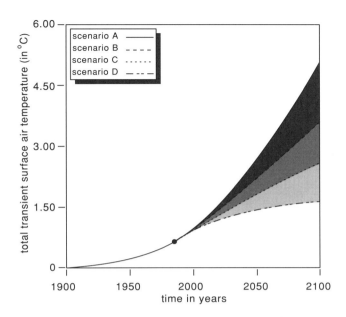

Figure 7.5: Global transient surface air temperature increase for the different scenarios, for $\Delta T_{2xCO_2} = 3°C$.

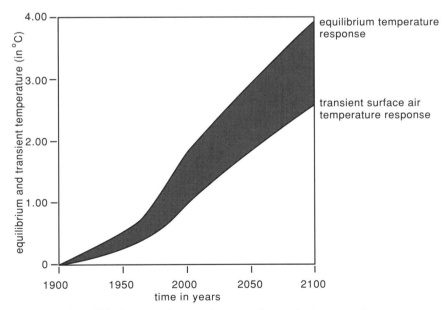

Figure 7.6: Equilibrium and transient surface air temperature response over the period 1900–1985, for $\Delta T_{2xCO_2} = 3°C$.

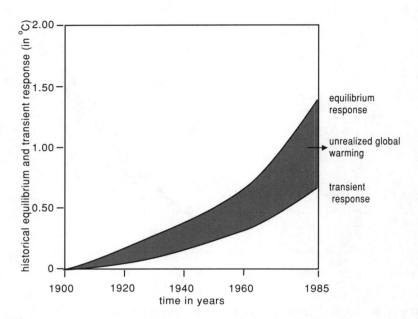

Figure 7.7: Equilibrium and transient surface air temperature response for a changed trends scenario, for $\Delta T_{2xCO_2} = 3^oC$.

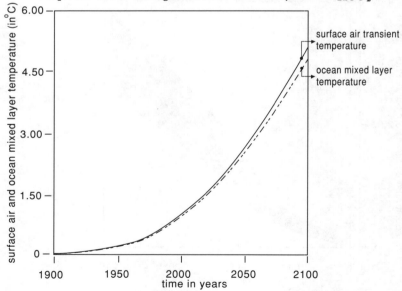

Figure 7.8: Surface air transient and ocean mixed layer transient temperature response.

7.3. RESULTS

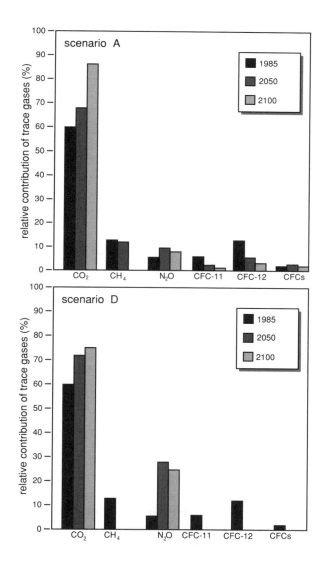

Figure 7.9: Relative contribution of trace gases to the greenhouse effect for scenario A and D.

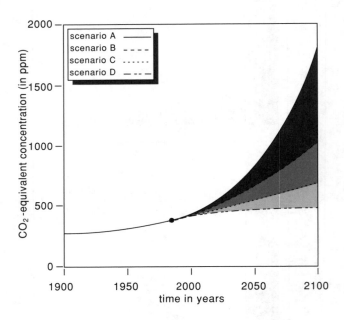

Figure 7.10: CO_2 equivalent concentration.

Finally a kind of sensitivity analysis has been performed. Firstly the diffusion coefficient is varied from 1000 to 8000 m^2/yr, with an average value of 4000 m^2/yr. The results are presented in Figure 7.11, giving the change in the surface air transient temperature response for the various values of the thermal diffusion coefficient. Figure 7.11 illustrates the importance of the diffusivity factor, and the sensitivity of the transient response to alterations in this factor.

7.3. RESULTS

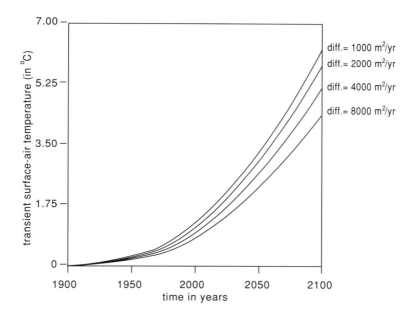

Figure 7.11: Transient surface air temperature response for different values of the diffusivity, 1000, 2000, 4000, and 8000 m^2/yr.

Secondly the thickness of the mixed ocean layer (Δz) is varied, from 25 meters to 150 meters. The results are depicted in Figure 7.12. The transient temperature response appears to be more sensitive to alternations of the depth mixed layer than for alternations of the diffusivity factor. However the uncertainty range for the diffusivity is larger than for the depth mixed ocean layer.

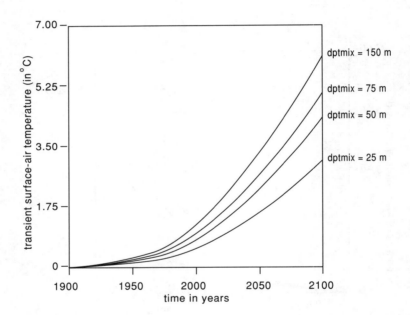

Figure 7.12: Transient surface air ocean temperature response for different values of the mixed ocean layer, 25, 50, 75, and 150 m.

Figure 7.13 gives the ocean response time or, equivalently, E-folding time (Schneider and Thompson, 1981). This E-folding time indicates the time to reach $1 - e^{-1}$ (or about 0.63) of the equilibrium temperature response for different values of the diffusivity coefficient. Assuming the standard value of the diffusivity, 4000 m^2/yr, yields an E-folding time about 50 years. Varying the diffusivity factor from 500 to 6000 m^2/yr yields a range in the E-folding time from 10 years to about 100 years.

7.3. RESULTS

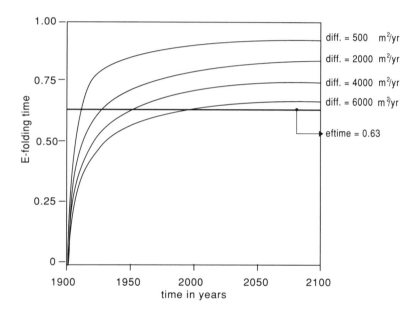

Figure 7.13: E-folding time at an instant CO_2 doubling, from 280 ppm to 560 ppm, for different values of the diffusivity: 500, 2000, 4000, and 6000 m^2/yr.

In Figure 7.14 the influence of the climate feedback factor, or, equivalently, the climate sensitivity to a doubling of the CO_2 concentration, is expressed. This figure shows that the unrealized warming depends strongly upon the climate feedback factor, or ΔT_{2xCO_2}. Assuming a value for ΔT_{2xCO_2} of 3.0 $°C$ or more, it can be concluded that only a decreasing fraction of the global warming has yet been realized.

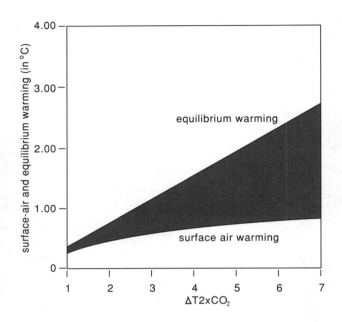

Figure 7.14: Surface air warming and the equilibrium warming for the period 1900–1985 as a function of climate sensitivity to a doubling of the CO_2 concentration, or, equivalently, the climate feedback factor.

7.4 Conclusions

The simple one-dimensional climate module within IMAGE cannot treat possibly important processes in any detail, including those that are determined by local or regional circumstances. Usually the different model parameters are turned within ranges found in the literature so as to mimic historical trends as far as these are known. Important aspects which are not or only very poorly incorporated in such a model are the treatment of clouds, ocean heat transport and circulation, and ocean-atmosphere feedbacks.

In spite of these shortcomings, these simple climate models are very useful. They can perform calculations over a long simulation period not

7.4. CONCLUSIONS

being feasible yet with three dimensional models. For policy evaluations, sensitivity analysis, and educational tools there is basically no alternative.

Calculations with IMAGE show that continuation of the recent trend of emissions of the relevant greenhouse gases leads to a rapid increase in the global mean temperature. Even the most restrictive forced trend scenario (D) leads to a 1.5 $°C$ temperature increase. The simple climate module within IMAGE indicates that more than half of the expected global warming effect has not yet been realized.

The CO_2 equivalent concentration, defined to combine the radiative perturbation of all greenhouse gases, doubles within 40 year for scenario A, whereas a doubling will not be reached for scenario D. Sensitivity analyses with the climate module show the sensitivity of the transient response to alternations of the climate feedback factor, the thickness of the mixed ocean layer and, to a lesser extent, to the diffusivity factor. Additional research is urgently needed to improve the knowledge which underlies the parameterizations in these simple climate models.

Chapter 8

The Sea Level Rise Module

Undoubtedly sea level rise is potentially one of the most threatening consequences of the greenhouse phenomenon. Nearly one-third of the world's population, including many of the world's largest cities, lives within 60 km of a coastline. Even a sea level rise of 1 meter would have a tremendous influence on habitation patterns, causing large-scale migrations of millions of people. Nevertheless the relation between a global temperature change and global sea level rise is not yet clear, and our knowledge of processes causing sea level rise is still deficient. Although on a large time scale geological processes are of crucial importance, on a relatively small time scale of centuries, three dominating causes of sea level rise are distinguished in the sea level rise module: thermal expansion of ocean water, melting of mountain glaciers, and the reaction of land ice. These causes associated with the greenhouse issue explain only a part of the basic sea level trend. Next to these anthropogenic causes, there are other factors which should be taken into account when studying the sea level rise. First, the internal variability of the climate system, regarded as all climatic fluctuations which are not related to anthropogenic activities or to changes in isolation of the earth (Oerlemans, 1982). Second, the systematic changes on long time scales. In this study only the latter are assumed to form the unexplained part of the basic trend, although it could be possible that climate variability is also related to the unexplained part of the basic trend.

8.1 Basic Trend

Sea level changes at any time and at any location is determined by both geological and climatic factors. Geological causes may have been responsible for a drop in sea level of about three hundred metres over the last eighty million years (Barth and Titus, 1984). But these geological events affecting sea level are generally slow and unlikely to accelerate. Generally the same holds for the climatic influence on sea level in the past. For the last two million years sea level rise and climate have changed together, in cyclic periods of about 100,000 years. During the last glacial era, sea level was approximately one hundred meters lower than today, while during the warm interglacial periods temperature and sea level have risen to approximately present levels (Barth and Titus, 1984).

In the last century sea level has risen by about 10–15 cm (Barnett, 1983 and 1985, Gornitz et al., 1982). The variations in these estimations arise from local effects. For the Netherlands and Belgium the sea level rise in the last century has been higher, about 20 cm. The global trend of 10–15 cm cannot yet be fully explained. Therefore in this study the basic or natural trend is assumed to be the unexplained part of the observed trend. Considering a trend of 20 cm per century for the Netherlands, the trend for the period 1900–1985 is about 17 cm. The model simulates for this period an anthropogenic sea level rise of 8.0 cm, consequently inducing a natural sea level trend of 9.0 cm. This induces a simulated natural trend of about 10.6 cm per century.

8.2 Thermal Expansion

Thermal expansion of the ocean is thought to be a major contributor to sea level rise. To calculate the thermal expansion effect, a simple box diffusion model is used, as described in Chapter 7. This approach, also used by Hoffman et al. (1983) and Oerlemans (1989), ignores the upwelling effect. Wigley and Raper (1987) use an upwelling diffusion model to calculate the thermal expansion effect. Here one single ocean column is used because, according to Oerlemans (1987), it follows, when splitting up the ocean into a warm and a cold column, the thermal expansion effects in both columns are similar. As described in Chapter 7, the diffusion equation is discretized, and numerically solved for a mixed ocean layer of 75 meters, and in 37 deep ocean layers of 25 meters. The diffusivity coefficients is kept constant, viz. 4000 m^2/yr. Using the negative exponential temperature profile as given by

Oerlemans (1989), each of the 38 ocean layers has its own thermal expansion coefficient and consequently its own contribution to thermal expansion. For each layer the sea level change due to thermal expansion can be described as:

$$SLR_{th}(i) = \int_0^{T(i)} [\Delta V(i)/V(i)] dz \tag{8.1}$$

and

$$\Delta V(i) = Td(i) * V(i) * ec(i) \tag{8.2}$$

with:
$SLR_{th}(i)$ = sea level rise by thermal expansion of layer i (in m)
$V(i)$ = volume of ocean layer i (in m^3)
$T(i)$ = thickness of ocean layer i (in m)
$Td(i)$ = transient temperature response in ocean layer i (in $°C$
$ec(i)$ = thermal expansion coefficient in layer i (in $°C^{-1}$)

At a certain time the total contribution of all layers is then:

$$SLR_{th} = \sum_{i=1}^{38} SLR_{th}(i) \tag{8.3}$$

Integrating these contributions for the whole simulation period yields the total sea level rise by thermal expansion.

8.3 Glaciers and Small Ice Caps

Although the ice mass of glaciers and small ice caps seems to be negligible in comparison with the immense ice masses of Antarctica and Greenland, their contribution to sea level rise may be substantial. Robin (1986) estimates the total area of mountain glaciers, all ice except for Greenland and Antarctica, as $0.54 * 10^6$ km^2 and $0.12 * 10^6$ km^3 water equivalent; this represents about 0.4% of total land ice, and corresponds to a potential sea level rise of about 33 cm. Oerlemans (1986b) estimates the total land ice mass at $0.6 * 10^6$ km^2 and $0.18 * 10^6$ km^3 water equivalent, corresponding to about 50 cm sea level equivalent.

Over the last hundred years there has been a world-wide retreat of the glaciers. The most comprehensive study in this field has been done by Meier (1984), who examined the mass balance of 13 glaciers. Meier (1984) estimates the contribution of glacier melting to sea level rise in the last hundred years to be 1 to 5 cm. Oerlemans (1988) argues that the greenhouse warming is responsible for about 50% of the glacier retreat over the last hundred years. Low volcanic activity is accountable for the remaining 50% of the glacier retreat.

Because local circumstances determine the specific character of each type of glacier, a global approach to the estimation of the entire melting of glaciers and small ice caps is rather precarious. So far, however, attempts to simulate the historic glacier retreat during the last hundred years, by using local data, have failed. Thus, for want of a more sophisticated approach, a simple global approach is chosen, according to Oerlemans (1989). Under the assumptions of a constant characteristic response time of glaciers, proportionality of melting rates to both temperature increase and remaining glacier volume, and an exponential decrease of glacier volume with surface temperature, the volume of glaciers and small ice caps can be described as follows:

$$Vglac(t) = Vglac(t-1)(t-1) + \int_{t-1}^{t} \left[\alpha * Ts * \left(Vglin * e^{-Ts/\beta} - Vglac(t) \right) \right] \quad (8.4)$$

with:
$Vglac$ = ice volume of glaciers at time t (in m sea level equivalent)
$Vglin$ = initial ice volume of glaciers; the starting simulation time is 1900, and the chosen initial glacier volume is 0.45 meter sea level equivalent
α = constant, which involves a characteristic response time of glaciers; is 0.05 $(yrK)^{-1}$
Ts = global transient surface temperature response (in $^\circ C$)
β = constant that determines the global temperature increase for which the ice volume becomes e^{-1} of the initial value; is 4.5 $^\circ C$

The constants are taken from Oerlemans (1989), while the global transient surface temperature increase is calculated with the climate module of IMAGE, described in Chapter 7.

8.4 Greenland Ice Cap

The Greenland ice cap covers an area of $1.8 * 10^6$ km^2, and a volume of $3 * 10^6$ km^3, corresponding with a sea level equivalent of about 7.5 meter (Oerlemans, 1989). Because of a lack of data it is not clear whether or not

8.4. GREENLAND ICE CAP

the Greenland ice cap is in its equilibrium state. In any case it is generally assumed that the Greenland ice cap is not far from its equilibrium state. In Rotmans (1986) three possibilities are distinguished, based on USDOE (1985c): firstly, according to the budget method (estimation of the total of accumulation, melting and calving), Greenland should be in equilibrium.

Accumulation	:	$+500 \pm 100 \ km^3$/year
Melting	:	$-295 \pm 100 \ km^3$/year
Calving	:	$-205 \pm 100 \ km^3$/year +
Net balance	:	0

Secondly, recent research indicates that the Greenland ice cap is in a disequilibrium state, varying from an increase of $+0.3 \ mm/yr$ to a decrease of $-0.7 \ mm/yr$.

Finally, extrapolating recent observations results in an ablation of 0.2 to 0.3 mm/yr and an accumulation of $-0.1 \ mm/yr$. The net effect, then, would be a thinning of $-0.1 \ mm/yr$.

In the model it will be assumed that in the initial phase, in 1900, the Greenland ice cap is in equilibrium (net balance is 0).

A very simple budget estimating method will be used to obtain a simple relationship between a global temperature increase and the change in mass balance of the Greenland ice cap. Assuming a global average temperature increase of 3.5 $°C$, this will elevate the height of the equilibrium line, by which the ablation surface area increases by about 25% and the melting rate increases by about 50% (Robin, 1986). Together these two factors would cause an increase in melting of about $-184 \ km^3$. The accumulation area would decrease with about 5%; accumulation rate stays constant or increases by 10% (USDOE, 1985c), yielding an extra accumulation of 0 or 47 km^3. The decrease in accumulation area and increase in accumulation rate would lead to an accumulation range of $-24 \ km^3$ to $+23 \ km^3$. Ultimately, for a 3.5 $°C$ temperature increase the net budget decrease would vary from $-208 \ km^3$ to $-161 \ km^3$. Supposing a linear relationship, the net effect per degree Celsius will vary from $-46 \ km^3$ to $-59 \ km^3$, corresponding with a sea level rise of 0.128 $mm/yr \ °C$ to 0.166 $mm/yr \ °C$, with an average value of about 0.147 $mm/yr \ °C$.

The contribution to sea level rise is modelled by multiplying this average melting rate of 0.147 $mm/yr \ °C$ to the simulated transient surface-air temperature increase.

In Rotmans (1986) the global temperature response is differentiated into several zones and into a summer and winter value, roughly based on GCM data. Because the melting process will occur only in the summer period, the average summer temperature increase is used, which, however, is chosen equal to the global average temperature increase.

Oerlemans (1989) summarizes several estimates which have been made for the change in mass balance of the Greenland ice cap per degree Celsius. He states that the contribution of the Greenland ice cap to sea level change can be estimated as 0.5 mm/yr per degree Celsius, with an uncertainty range of about 50%. This is rather high compared to the value found here of 0.147 mm/yr per degree Celsius. This is mainly caused by the accumulation rate, which is a maximum of 10% in this analysis, and which is being kept constant in the analysis, leading to 0.5 mm/yr per degree Celsius.

8.5 Antarctic Ice Cap

The Antarctic ice sheet contains $11.97*10^6$ km^2, a volume of $29.33*10^6$ km^3, which is more than 90% of the total amount of land ice, and corresponds to about a 65 meter sea level equivalent. As for the Greenland ice sheet, it is generally supposed that the Antarctic ice sheet is not far from its equilibrium state. There is still an uncertainty range of about 20%. According to a simple budget method, the Antarctic ice sheet should be slightly increasing:

Accumulation	:	$+2000$ km^3/year
Ice flow	:	-1600 to -2000 km^3/year $+$
Net balance	:	0 to $+400$ km^3/year

On the Antarctic ice cap the ice flow is primarily caused by calving, while the melting is negligible, because of the extremely cold climate. In case of a temperature increase the accumulation is expected to increase, causing a drop in sea level. To calculate coarsely the effect of a temperature rise on the Antarctic mass balance, again a simple budget estimation is used (Rotmans, 1986).

Assuming once more a global average temperature increase of 3.5 $°C$, this may lead to an increase in accumulation by about 10 to 25%, corresponding to about $+200$ to $+500$ km^3/year. On the other hand, the ablation rate will double over twice the ablation area, implying a quadrupling of the ablation. To determine the net balance effect in consequence of such a global temperature rise, several options can be chosen (Rotmans, 1986). One with an initial net balance of zero, or one with a net balance of +20%, or a 10% or 25% accumulation rate increase, etc. Taking account of these various options, the range, expressed in mm/yr per degree Celsius, varies from -0.1088 mm/yr per degree Celsius to -0.3648 mm/yr per degree Celsius. Selecting an average net budget increasing value of -0.237 mm/yr per degree Celsius, this is multiplied by the simulated transient surface-air temperature increase in summer, which value is parameterized as 1.5 times the global transient surface-air response. The resulting net budget increasing value is then -0.36 mm/year per degree Celsius.

Oerlemans (1989) estimates the contribution to sea level change in two different ways. On the one hand, by using a model for the Antarctic ice sheet (Oerlemans, 1982), resulting in an estimation of -0.5 mm/yr per degree Celsius, and on the other by directly considering the mass balance effect, leading to a -0.43 mm/year per degree Celsius estimate. This does not deviate too far from the value used in this study of 0.36 $mm/yr\ ^\circ C$. The possible disintegration of the West Antarctic Ice Sheet (WAIS) is not taken into account in this analysis, because it seems unlikely that this disintegration will already occur in the next century. Even if this happens the effects in the coming hundred years will be small (National Health Council, 1986). Although there is ample qualitative knowledge about the ice sheet system, it is hard to make quantitative projections. Various attempts have been made to model the West Antarctic ice sheet system (Thomas and Bentley, 1978, Van der Veen, 1986). Oerlemans (1982, and 1989) argues that the earlier modelling estimates were too high, and estimates the contribution of the West Antarctic Ice Sheet at about 0.1 mm/yr for the next century. However this quantification is surrounded by large uncertainties, and is consequently ignored in this study.

8.6 Uncertainties

The uncertainties with respect to future sea level rise projections are very large. The uncertainties in the various separate contributions to future sea level rise are about 50% for the changes on the Greenland and Antarctic ice cap, about 50% for the glacier melting, and approximately 30% for the thermal expansion effect. The uncertainties of the individual contributions are assumed to be independent of each other. Concerning the total or accumulated uncertainty, Oerlemans (1989) argues that 40% of this is caused by uncertainty in climate modelling, and the remaining 60% is due to lack of data and inadequacy of the models that explain sea level rise.

8.7 Sea Level Rise Potential

Effective targets are needed in order to develop environmental long-term goals with respect to the effects of climate change. Next to global temperature increase, sea level rise might be used as a long-term target for climate change, from which concerted emission control policies can be derived. Sea level rise should be taken into consideration, since many valuable ecological and economic areas are in coastal regions, and even a moderate sea level rise could have a potentially disastrous effect.

When considering sea level rise as a long-term target of climate change, a tolerable sea level rate should be defined. According to Gornitz et al. (1982) the sea level trend over the past 6000 years is 2 cm per century, which is very small as compared to long-term trends of 1 meter per century at times of continental ice sheet growth or decay. As mentioned above, however, the global trend over the last hundred years is about 10–15 cm. This interval of 10 to 15 cm per century could be taken as a reference target value for sea level rise for the coming century. Another possibility is to relate the sea level rate to global temperature increase recommendations, as was done at the Villach/Bellagio workshop (Bolin et al., 1986, Jäger 1988). There a reference value of 0.1 $°C$ per decade was suggested, a rate of change at which ecosystems might be able to adapt effectively to climate change. Using this global temperature increase scenario of 0.1 $°C$ per decade as input scenario for IMAGE, a global sea level rise of about 20 cm per century is calculated, whereas for the Netherlands a sea level rise of 25 cm would be expected. Thus, a global temperature increase target of 0.1 $°C$ per decade would keep pace with a global sea level rise target of 20 cm per century (25 for the Netherlands).

If the control of sea level change is to be used as a long term goal, it should be possible to derive emission pathways from this goal. Therefore an index to compare the sea level rise effect of greenhouse gas emissions is needed. By analogy with the Temperature Increasing Potential (TIP), which is in turn a greenhouse counterpart to the Ozone Depleting Potential (ODP), here the concept of a Sea Level Rise Potential (SRP) is introduced. IMAGE is used to obtain a relationship between an emission and its associated effect on sea level. The complexity of the processes involved seems to make an analytical approach, as used for determining the TIP concept, difficult.

In view of the strong resemblance between the TIP and the SRP, calculation of both potentials will be based on the same methodology. This modelling approach is described extensively in Chapter 11, and in Rotmans and Den Elzen (1990). The modelling method will be explained briefly below.

Again, the SRP is defined as the sea level rise effect of 1 or 10^{-3} Gt emission of a particular trace gas relative to CO_2. For that, stabilizing scenarios have been developed for each trace gas, resulting in steady-state concentrations in the second half of the next century. Starting these scenarios in 1985, in 1986 for each trace gas a one year emission impulse of 1 or 10^{-3} Gt is generated, and added to such a stabilization scenario. Then in each case two stabilization scenarios are compared in pairs, one with and one without an emission impulse. In this way, for each greenhouse gas, two sea level rise responses are simulated, again with and without an emission impulse. By subtracting these two responses, the influence of the scenario choice is reduced, yielding the net sea level rise effect. These net contributions are integrated over the whole simulation period. Dividing this sea level rise integration of CH_4, N_2O, CFC-11, and CFC-12 by that of CO_2 gives the SRP. Again, because CO_2 has no specific atmospheric residence time, for CO_2 two time horizons have been taken into account, 100 and 500 years, respectively. Also, the same parametrizations have been used as in the TIP modelling approach (Rotmans and Den Elzen, 1990).

As opposed to the equilibrium states as used in the TIP concept, the complex nature of the forcing processes in simulating the SRP compels the use of transient responses.

8.8 Results

The total sea level rise is calculated for the four scenarios. The resulting sea level rises are presented in Figure 8.1. The sea level rise range varies from about 0.95 meter for the unrestricted scenario (A) to about 0.45 meter for the forced trends scenario (D). Considering the large uncertainties mentioned above, it appears that for the changed trend scenario (C) the outcomes of the unrestricted trend and the forced trend scenarios are approximately the upper and lower bounds, respectively, of such a simple uncertainty analysis.

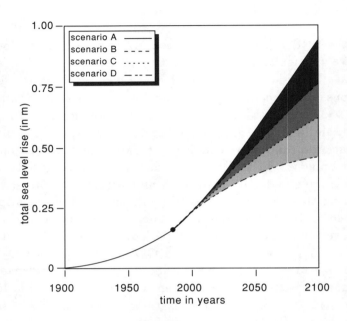

Figure 8.1: Total sea level rise for the different scenarios.

Figure 8.2 gives the different components of the sea level rise for the unrestricted trends scenario of about 0.95 meter. Clearly, the thermal expansion contribution dominates. Next to thermal expansion, melting of glaciers and small ice caps play an important role. The melting of the Greenland ice cap and the accumulation of Antarctica are of minor importance in this simulation.

8.8. RESULTS

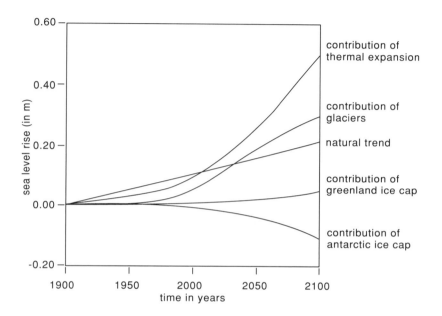

Figure 8.2: Different components of the total sea level rise for the unrestricted trends scenario (A).

Figure 8.3 contains the relative contributions of the different components to sea level rise. These simulation results show primarily an increasing share of thermal expansion and glacier melting, and a decreasing natural sea level rise component. For 1985 the total simulated sea level rise is 17 cm, of which 8 cm has an anthropogenic nature and the remaining 9 cm forms the basic trend. This is in line with the 'explained' sea level rise of about 9.5 cm over the last 150 years found by Oerlemans (1989). The major contributor is the thermal expansion of about 6 cm, which is fairly high as compared to Oerlemans' estimate of 5 cm, and the estimated range of 2–5 cm found by Wigley and Raper (1987) for the thermal expansion effect over the last hundred years. The reason is probably the use of a constant diffusion coefficient and larger expansion coefficients.

Another important contributor are the glaciers and small ice caps, with 2.3 cm. Oerlemans gives about 3.5 cm, while Meier (1984) gives a range of

1 to 5 *cm* over the past hundred years.

Generally, the simulated sea level rise for 1985, as well as the separate contributions of different components, are reasonably in line with results given in the literature.

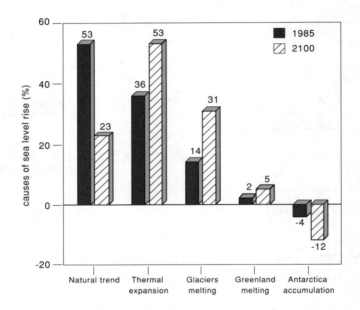

Figure 8.3: Relative contribution of the different components to sea level rise for the years 1985 and 2100; the latter for the unrestricted trends scenario (A).

The sea Level Rise Potentials (SRPs) of the different greenhouse gases are represented in Table 8.1.

SRP value	CO_2 time horizon is 100 years	CO_2 time horizon is 500 years
CO_2	1	1
CH_4	15	6
N_2O	275	330
CFC-11	3780	2560
CFC-12	8250	5000

Table 8.1 SRP calculated with IMAGE for different time horizons.

Comparing these SRP values with the TIP values presented in Chapter 11, and in Rotmans and Den Elzen (1990), shows that for a CO_2 time horizon of 100 years the SRP values are slightly lower than the TIP values. For a CO_2 time horizon of 500 years, however, the SRP values are considerably higher. This can be explained by the fact that the SRP is based on transient states, whereas the TIP indicates an equilibrium phase. The transient response induces a delay effect which causes a shift towards the time horizon of 500 years. For instance the SRP for N_2O is higher for the 500 years time horizon than for the 100 years horizon.

8.9 Conclusion

Simulation experiments with IMAGE show a sea level rise range of about 0.45 to 0.95 meter for the next century. The simulated increase for the next hundred years falls within the interval of 0.28 to 0.78 meter, which means an increrase of about 2 to 5 times that of the past centuries. The dominant causes of this sea level rise will be the thermal expansion of the ocean, and the melting of glaciers and small ice caps. For the next hundred years changes in the mass of polar ice sheets may have a significant effect on sea level, but will be minor components compared to expansion and glaciers.

Notwithstanding the large uncertainties with which the processes that determine a sea level rise are surrounded, it seems worth the effort to develop sea level rise targets. The IMAGE model has been used to develop Sea Level Rise Potentials (SRPs) for the greenhouse gases, analogous to the

Temperature Increasing Potentials (TIPs) described in Chapter 11. SRPs can serve as a useful tool for setting long term goals in environmental policy. SRPs could be used to estimate greenhouse gas emissions associated with a specific sea level rise target. The results presented here show that it is possible to derive SRPs for the various greenhouse gases.

The SRPs are not identical to the TIPs because the SRPs are based on transient responses, whereas TIPs are calculated for the equilibrium response. As a result, the SRP estimates are relatively more uncertain than the TIP estimates.

Chapter 9

Socio-Economic Impact

9.1 Introduction

IMAGE contains a separate module for roughly estimating the socio-economic impact of global temperature and sea level rise on Dutch society. The module is based on a tentative socio-economic study, not pretending to give a comprehensive overview of the consequences for the Netherlands. Sea level rise is a major issue in the Netherlands, because of the country's vulnerability to rising sea level, which has resulted in the so-called Delta flood protection plan after the last tragic flood in 1953 (2000 people drowned). The socio-economic impact module includes most relevant aspects that can be quantified for the Netherlands. The most important impacts considered relate to the protection of coastal area (by dikes and dunes) and the adaptations required for the water management. Four consistent sets of scenarios have been worked out, based on differences in economic growth, energy use, international environmental measures etc, which are described in Chapter 2. Given these scenarios estimates are made of the costs of coastal defence and water management in the Netherlands as a result of adaptation to impacts of regional climate change and sea level rise. For other social sectors, such as agriculture and energy use, only tentative conclusions are drawn.

Ecological effects have not been considered yet but, on a European scale, some will be added to the existing model in the near future. In view of the many uncertainties, the results from this socio-economic simulation model give only indicative ranges of costs for the different sectors.

9.2 General Model Description

As mentioned before the modelling framework used is the IMAGE model, which has been described extensively in previous chapters. In this study all socio-economic consequences are based upon the four different scenarios, of which an elaborate description is given in Chapter 2 and in Rotmans et al. (1990a). The highest scenario, A: *unrestricted* trends, assumes a continuation of economic growth, not limited by environmental constraints. Scenario B: *reduced* trends, supposes the implementation of environmental measures presently being considered to control other environmental problems like acidification. Scenario C: *changed* trends, assumes the enforcement of stricter environmental control. Finally, scenario D: *forced* trends, assesses the possibilities of maximum efforts efforts towards global sustainable development. World population growth, a factor that is assumed not to be influenced by greenhouse policies, reaches 10.8 billion in 2100 in all scenarios.

Based on these four scenarios IMAGE calculates global temperature increase and sea level rise (Figures 9.1 and 9.2), providing the input for the socio-economic impact model. One of the missing links of the model is the regional shift in the main hydrological pattern, direcly influencing the runoff of the great rivers Rhine and Meuse and thus the country's water balance.

The emissions of trace gases are input for the calculation of several global phenomena such as trace gas concentrations, temperature increase and sea level rise. The latter two are input for the socio-economic impacts modules, in which the costs and benefits of various socio-economic consequences for the Netherlands are calculated. The separate model on the socio-economic impacts of the greenhouse effect for the Netherlands again consists of independent, interlinked modules, and is based on both an extensive study of the literature and knowledge transfer resulting from a close cooperation with specialized Dutch experts of Rijkswaterstaat Netherlands, Delft Hydraulics, Dutch Organization for Applied Scientific Research (TNO) etc. Figure 9.3 gives the modular build-up of the socio-economic impact model.

Simple dynamic modules have been developed, which will be described in the next section, reflecting the complicated mechanisms in the fields of coastal defence and water management, resulting in dynamic relationships between input (climate change, sea level rise) and output (damage or profit expressed in terms of money).

9.2. GENERAL MODEL DESCRIPTION

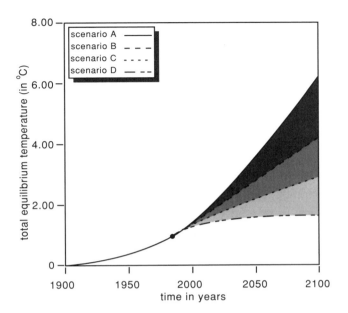

Figure 9.1: Global mean equilibrium temperature increase.

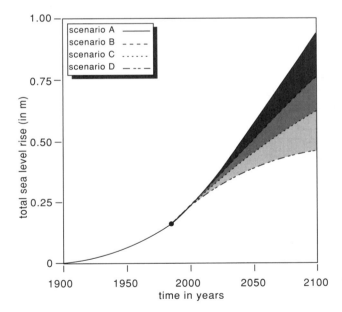

Figure 9.2: Global mean sea level rise.

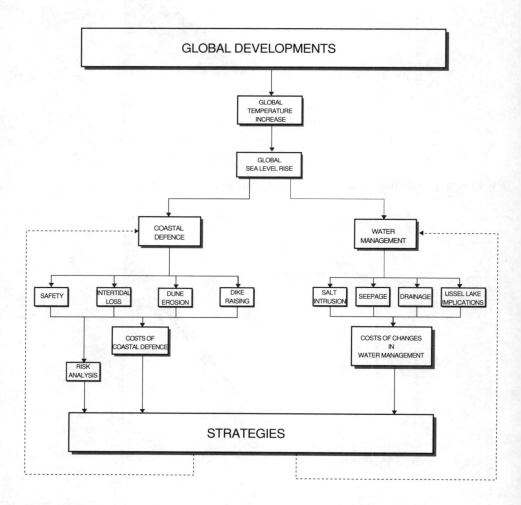

Figure 9.3: Socio-Economic Impact Model for the Netherlands.

9.2. GENERAL MODEL DESCRIPTION

The impact modules have a simulation period of 110 years, from 1990 to 2100, unlike the simulation period of 200 years for the global phenomena. For the agricultural and energy sector only tentative quantifications have been made, due to large uncertainties with respect to regional climate changes, in particular the changes in the hydrological cycle.

In Figures 9.4a, 4b and 4c the spatial segmentation of the Netherlands is given for the coastal defence and water management (seepage and drainage). This picture shows three kinds of coastal defence systems, dikes, dunes and intertidal zones. The dikes are segmented into three parts, the Delta area, the Western and the Northern part. The dunes are mainly concentrated in the Western part and the Frisian Islands. The main intertidal areas for the Netherlands are the Wadden area and the Delta area. For the Dutch coastal zones the optimal economic dike height is calculated as well as the erosion of dune areas and losses of intertidal areas. However the latter is done in a very schematic way, considering the barely known changes in the morphology of the intertidal zones, which are mainly determined by the increase or decrease of the tidal volume in relation to the change of the cross profile of the tidal inlet (de Ronde, 1988). Another difficulty relates to the economic value of intertidal areas, so the socio-economic consequences for the intertidal zones are not taken into account.

With regard to the water management sector various areas have been differentiated, 22 seepage areas, 7 drainage areas, the New Waterway area for salt intrusion and the IJssel lake. Based on this zoning both the principal financial consequences and the various alternative countermeasures for the water management sector are set out.

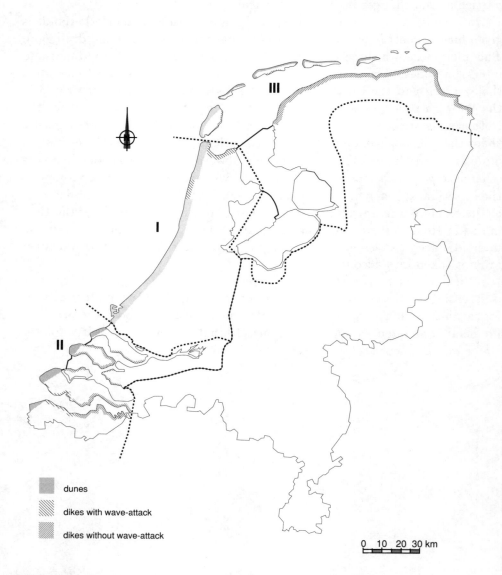

Figure 9.4a: Coastal Defence system of the Netherlands.

9.2. GENERAL MODEL DESCRIPTION

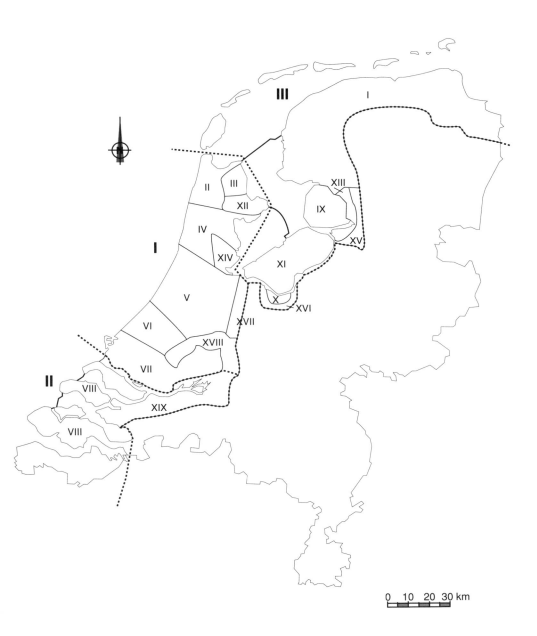

Figure 9.4b: Seepage regions for the Netherlands.

170   CHAPTER 9. SOCIO-ECONOMIC IMPACT

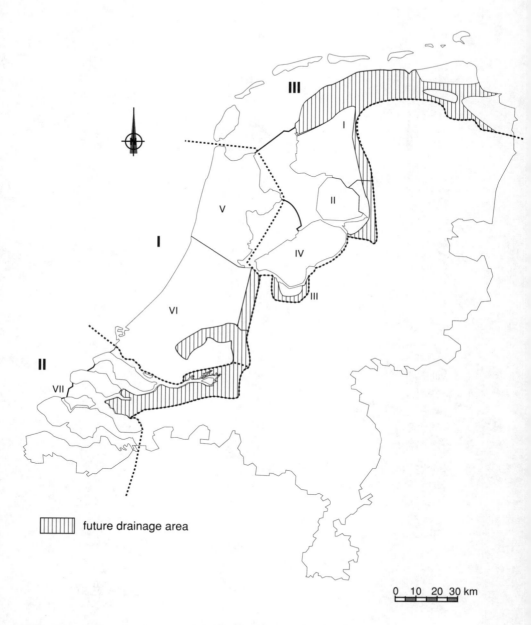

Figure 9.4c: Drainage regions for the Netherlands.

9.3 Quantification of Impacts for Various Sectors

9.3.1 Introduction

The Netherlands have been struggling with the sea level for hundreds of years already, and this has resulted in a highly organized protective system of dikes and dunes. A sea level rise induced by climate change would affect the structures of this protective system. A sea level rise is therefore of crucial importance to the Netherlands. Because it is not yet possible to quantify ecological and geomorphological impacts, in this study only the economic and societal implications of sea level rise are modelled for the Netherlands. The economic impacts relate to the protection of coastal area by dike raising and dune strengthening, and adaptations required for water management. With respect to dike raising several mathematical models have been developed. One model considers dike raising as an economic decision problem, taking into account the safety levels, dike raising costs, and the expected damage costs in case of an inundation. Another model primarily focuses on stabilizing the present safety level. Concerning the sandy coast protection our model depends heavily upon more detailed models of Rijkswaterstaat.

Next to economic impacts of sea level rise in socio-economic terms are also treated, such as loss of intertidal areas, salt intrusion, seepage, drainage problems, etc. One of the missing links in our socio-economic modules is the regional shift in the main hydrological pattern, directly influencing the runoff of the great rivers Rhine and Meuse and so the country's water balance.

In the following sections the background and mathematical dynamics of the modules concerning coastal defence, both fixed coast and sandy coast, and water management are presented. It should be noted that all sums of money have been expressed in terms of Dutch guilders (Dfl: is approximately 0.5 US Dollar).

9.3.2 Coastal Defence

The Netherlands' coast is half protected by dunes and half by dikes (see Figure 9.4a). Concerning the dikes, two types of dikes are distinguished, dikes with wave attack and dikes without wave attack, which differ in average height. Dikes with wave attack have a height between 10 and 15 meters above N.A.L. (Normal Amsterdam Level), whereas dikes without wave attack are lower, from 5 to 7 meters above N.A.L. This difference has to do with the situation of the dikes. Dikes with wave attack are situated along the Dutch coast in direct contact with the open North Sea, and must be able to resist a storm surge of +5 meter N.A.L. and its attendant heavy wave attacks. These heavy wave attacks necessitate an extra dike height of 5 till 10 meter, on top of the already required +5 meter N.A.L. for the storm surges, resulting in a total dike height of 10 to 15 meters N.A.L. By contrast, the dikes without wave attack are situated in Zeeland and along the Northern coast, having no direct contact with the North Sea, see Figure 9.4a. Therefore for these dikes a dike height of 5 to 7 meters will suffice. As for the dunes no distinction has been made with respect to dune height, but a segmentation into 13 dune segments has been used, based mainly on morphological properties (Den Elzen and Rotmans, 1988).

Assuming that the Dutch coastal defence system will only be extended and reinforced, but not fundamentally altered, for these two kinds of coastal defence systems an attempt has been made to model the consequences of an accelerated sea level rise. When raising the dikes, two additional features should be taken into account. First the additional sea level rise of 5 centimeters due to the high tide level. Consequently, maintaining the present safety level implies an additional dike raising of 0.05 meter (Eversdijk, 1989). Second, the extra strength of the higher waves requires an extra raising of the dikes with wave attack, represented by a correction factor.

Next the following aspects are discussed: safety, dike raising, and dune strengthening.

Safety

Safety is defined here as the protection against flooding, which can be expressed in terms of the frequency of overtopping the flood protection system. After the disastrous flood in February 1953 in Zeeland, by which most of the Delta area in the Netherlands was flooded, almost all dikes along the Dutch coast were raised, now enabling them to resist a storm surge of +5 meter N.A.L. The present theoretical chance of inundation, the so-called "Delta norm" is 0.0001 for Central Holland, being a risk of inundation once every 10,000 years, and 0.00025 for the other threatened areas in the Netherlands, Zeeland and the Northern part. The last big storm of 1953 reached +3.80 meter N.A.L. and had a statistical chance of inundation of 0.0045 (Delta Commission, 1960).

The relationship between frequency of overtopping and sea level rise is assumed to be exponential, based on historical sea level data (Van Dantzig, 1956, De Jong, 1985, Rijkswaterstaat, 1989, De Ronde, 1988, Delta Com-

9.3. QUANTIFICATION OF IMPACTS FOR VARIOUS SECTORS

mission, 1960 and Vrijling, 1985):

$$P(h \geq H)_t(i) = P(h \geq H)_0(i) * e^{ln(10)*\alpha*\Delta H_t(i)} \qquad (9.1)$$

with:

$P(h \geq H)_t(i)$	=	probability of exceeding of dike height H at time t in area i, $i = 1, 2, 3$
H	=	dike height above mean sea level (0 m N.A.L.) (m)
α	=	gradient of flooding frequencies of storm surges (= 1.5, Wind, 1987)
$P(h \geq H)_0(i)$	=	the present safety norm (equals to 0.0001 for area 1 and to 0.00025 for area 2 and 3)
$\Delta H_t(i)$	=	$SLR_t + \psi + CRFC - DR_t(i)$
SLR_t	=	sea level rise at time t (m)
ψ	=	0.05, as a consequence of an extra rise of 0.05 meter for the high tide, when sea level rises (m)
$CRFC$	=	correction factor for dike raising, caused by extra wave height in case of dikes with wave attacks (m)
	=	$\beta * SLR_{2100}$
β	=	0.6, for dikes with wave attack (De Ronde and De Vogel, 1989)
	=	0.0, for dikes without wave attack
$DR_t(i)$	=	total dike rise at time t for area i (m)

Practically, this means that reducing the present safety norm, or equivalently, Delta norm, of 0.0001 to a tenth, involves a necessary dike raising of about 0.70 meter.

Dike Raising

An accelerating sea level rise will cause a decrease in safety, according to Equation (9.1), meaning numbers of people and goods at risk. Countering this threat requires a raising of the dikes. Because it seems impossible to predict both sea level rise and the pace of it, we assume dike raising will be carried out stepwise. Construction in one step would mean an unnecessarily high financial and material risk. Based on this assumption, the total dike raising for the whole period of 1990 till 2100 is calculated. Subsequently an algorithm is applied which generates the time pathway of a stepwise dike raising for the whole period till 2100, using the already calculated total dike raising. This algorithm is described in detail in Den Elzen and Rotmans (1988).

CHAPTER 9. SOCIO-ECONOMIC IMPACT

As mentioned before, the construction of a total dike raising operation depends upon the type of dike considered. Next to adaptation to sea level rise itself, additional standard raisings are necessary due to both high tide level and higher waves, the latter only for dikes with wave attack, in order to maintain the present Delta norm.

The problem of dike raising is here considered as an economic decision problem, where the criterion for dike raising is related to the value of the people and materials at risk. This implies that the present safety norm will be no longer given in advance as 0.0001, but should be calculated based on this economic analysis. The economic decision problem is the calculation of an optimal dike raising at minimal costs for the period 1990–2100. This results in the following total dike raising equation:

$$DR(i)_{2100} = OPDR(i) + \psi + CRFC \qquad (9.2)$$

with:
$DR(i)_{2100}$ = total dike raising in the year 2100 in area $i = 1, 2, 3$ (m)
$OPDR(i)$ = optimal dike raising in area i (m)
ψ = 0.05
$CRFC$ = $\beta *$ SLR$_{2100}$ (m)
SLR_{2100} = sea level rise at the end of the simulation period, in 2100 (m)
β = 0.6, for dikes with wave attack (De Ronde and De Vogel, 1989)
  = 0.0, for dikes without wave attack

The goal function to be minimized in this analysis, the total costs over the period 1990 to 2100, consists of the sum of the costs of dike raising, and the capitalized damage expectation of the material and human losses when dike breach occurs. This goal function is expressed in Equation (9.3) (Den Elzen and Rotmans, 1988 and Den Elzen et al., 1990). The costs of dike raising are assumed to be linear with dike raising; this means that if dike raising equals x meter, the costs of dike raising will be proportional to the product of the marginal costs of dike raising per meter raising per kilometer dike, and the factor x. To carry out this economic analysis in a sound way all economic monetary units (in Dfl) in the goal function (9.3) have to be discounted to the year 1990.

$$l(i) * [I_0(i) + k(i) * OPDR(i)]$$
$$+ \sum_{t=1990}^{2100} (1 - \delta/100)^{t-1990} * W_0(i) * P(h \geq H)_0(i) *$$

9.3. QUANTIFICATION OF IMPACTS FOR VARIOUS SECTORS

$$e^{\alpha*ln(10)*(SLR_t-OPDR(i))} \tag{9.3}$$

with:
- $OPDR(i)$ = optimal raising of the design level in area i (m)
- $I_0(i)$ = initial costs of dike raising for one kilometer dike (Dfl/km)
- $l(i)$ = length of dikes in area i (km)
- $k(i)$ = marginal costs of dike raising in area i per meter raising per kilometer dike (Dfl/m.km)
- δ = reduced rate of interest (real interest minus inflation) (in %)
- $p(h \geq H)_0(i)$ = 0.0001, for area 1
  = 0.00025, for area 2 and 3
- $W_0(i)$ = economic value of area i at risk in 1990 (Dfl)
  = $p_d * N(i) * V_0(i) + M_0(i)$
- $M_0(i)$ = capitalized value of the protected goods, including firms, cattle, etc. for area i (Dfl)
- p_d = the probability of dying given an inundation (see Table 9.1, Vrijling, 1985)
- $N(i)$ = number of protected people in area i
- $V_0(i)$ = capitalized value of one human being in 1990 (see Table 9.1) (Dfl)
- α = gradient of flooding frequencies (see Equation (9.1))
- SLR_t = sea level rise at time t (m)

This reaches a minimum for the optimal dike raising:

$$OPDR(i) = ln((ln(10) * \alpha * W_0(i) * P(h \geq H)_0(i) * \sum_{t=1990}^{2100} (1-\delta/100)^{t-1990} * e^{ln(10)*\alpha*SLR_t})/k(i)*l(i))/\alpha \tag{9.4}$$

The values of all parameters used are listed in Table 9.1.

For calculating the total costs for dike raising, the inner dikes in Zeeland are also taken into account. The raising of the design level for those dikes is supposed to be equal to the raising of the design level throughout Zeeland. The algorithm generating the stepwise execution of dike raising is described in detail in Den Elzen and Rotmans (1988).

In Den Elzen and Rotmans (1990) a second strategy, based on maintaining the present safety norm is discussed. The resulting dike raisings for both methods are presented and compared with each other.

Dune Strengthening

An accelerated sea level rise will increase present erosion, which is about 6.5 million m^3 yearly (Rijkswaterstaat, 1989, Stive, 1989, Beafort et al., 1989), and will cause a retreat of the whole coastline. This would jeopardize the safety of many people and material goods in the areas lying behind the dunes. Therefore the dunes will have to be strengthened. With the help of computer simulation models developed by Rijkswaterstaat, (Dillingh et al., 1984), the dune erosion can be simulated. In Knoester et al. (1989) assessments have been made for the coastal retreat for a sea level rise in 2090 of 0.2 meter, and 0.6 meter, respectively. In view of all uncertainties concerning coastal morphological processes, a linear relationship between the coastal retreat and the rate of the sea level rise is supposed. Thirteen dune areas are distinguished, and for each area the coastal retreat is calculated at any time, interpolating with linear regression the already known reference points at a sea level rise of 0.2 and 0.6 meter:

$$CR_t(i) = a(i) * RSLR_t + b(i) \tag{9.5}$$

with:
$CR_t(i)$ = coastal retreat at time t for area i, $i = 1, 2, \ldots, 13$ ()
$a(i), b(i)$ = coefficients for area i
$RSLR_t$ = rate of sea level rise at time t (m)

The present policy on dune erosion implies concentration on taking measures only in case of threats to safety, nature, water supply or recreation. A further retreat of the dunes and the coastline is socially unacceptable and moreover too expensive in the long run. So, for large parts of the Dutch coast, full compensation of erosion such as sand depletion is necessary. The costs of a policy to arrest dune erosion are simulated on the basis of the calculated dune erosion. These costs are composed of the present costs (about 30 million Dfl yearly) and the costs caused by extra dune erosion (about 5 Dfl per m^3 erosion (Vrijling, 1985)).

In Knoester et al. (1989) the relation between coastal retreat and yearly dune maintenance costs, as well as the relation between coastal retreat and costs for dune erosion is given. The total costs for the dunes can then be expressed by:

$$DC_t = \sum_{i=1}^{13} m * CR_t(i) + e * CR_t(i) \tag{9.6}$$

9.3. QUANTIFICATION OF IMPACTS FOR VARIOUS SECTORS

with:
DC_t = total costs for dunes at time t (Dfl)
m = coefficient caused by dune maintenance costs (Dfl/m)
e = coefficient caused by dune erosion costs (Dfl/m)

	area 1 Central Holland	area 2 Zeeland (+inner Zeeland)	area 3 Friesland and Groningen
gradient of frequencies (α)	1.5	1.5	1.5
present safety norm $p(h \geq H)_0$	0.0001	0.00025	0.00025
Initial costs of dike raising (I_0) (Dfl/m)	$3 \cdot 10^6$	$3 \cdot 10^6$	$3 \cdot 10^6$
marginal costs of dike raising (k) (Dfl/m.km)	$4 \cdot 10^6$	$4 \cdot 10^6$	$4 \cdot 10^6$
reduced rate of interest	2	2	2
economic value of materials at risk (Dfl)	$1.418 \cdot 10^{12}$	$2.67 \cdot 10^{11}$	$2.85 \cdot 10^{11}$
chance of dying at inundation	0.01	0.01	0.01
capitalized value of a human being (Dfl)	$5 \cdot 10^6$	$5 \cdot 10^6$	$5 \cdot 10^6$
number of people (in 1985)	$6.05 \cdot 10^6$	$0.85 \cdot 10^6$	$1.3 \cdot 10^6$
length of dikes (km) with wave attack without wave attack	45	90 (370)	15 150

Table 9.1: The values of the parameters used in the socio-economic model. The economic values are based on projections of the Central Planning Bureau (1984).

9.3.3 Water Management and Water Supply

For the water management and water supply sectors the consequences of an accelerated sea level rise and an intensification of the hydrological cycle have been examined and modelled. Four different conceivable problems are involved: salt intrusion, seepage, drainage, and the maintenance of the IJssel lake.

Salt Intrusion

In the Netherlands salt intrusion mainly occurs in two branchings of the New Waterway: Old Meuse and New Meuse - Dutch IJssel (see Figure 9.4). The inland salt load due to salt intrusion depends, according to a method developed by Bruggeman (1988a), on the slope and water depth of the New Waterway and discharge of the Rhine and Meuse. In this method the sedimentation pattern of the New Waterway is assumed to remain unchanged. Another influential factor in the calculations is the discharge of the New Waterway. In light of the uncertainties about the regional pattern of an impending climate change it is not yet possible to estimate this future discharge. Therefore it is assumed to be constant in time. In case of a sea level rise the salt tongue intrudes further inland, and it passes a fixed point in the Old Meuse or in the Dutch IJssel more frequently. This results in an increasing salinity in the Westland (an area near Rotterdam), and causes great agricultural damage.

Moreover, the increasing salt intrusion threatens several water-collection areas. The total costs incurred because of salt intrusion are much higher than costs of intrusion-preventing actions (Pulles, 1985).

Seepage

Seepage is the constant underground current of salt water in the direction of low-lying areas as a consequence of the head difference between sea level rise and large areas situated below sea level behind the dunes (see Figure 9.4). The inland salt load caused by seepage depends on the resistance of the geological beds in the dunes, the difference between sea level and polder level and precipitation (Bruggeman, 1988b). Sea level rise occasions not only an increase in the difference between sea level and polder level, but also an enlargement of regions below sea level, together leading to an increasing seepage in the regions behind the dunes. This yields the following expression

9.3. QUANTIFICATION OF IMPACTS FOR VARIOUS SECTORS

for the total salt load due to seepage in the 22 specific seepage regions:

$$Q_t(i) = L * (G_1(i) * (H(i) + SLR_t) + \alpha * G_2(i) * N) \tag{9.7}$$

with:
$Q_t(i)$ = salt load flux in area i at time t (in m^3/day)
$i = 1, \ldots, 22$ (22 seepage regions)
L = length of the coast line (in m)
$G_1, G_2(i)$ = geological constants of the dunes in area i (in m/day)
$H(i)$ = difference between the sea and polder level in area i (in m)
SLR_t = global sea level rise at time t (in m)
N = yearly mean precipitation (in m)
α = geohydrological correction factor

The primary effect of seepage is great agricultural damage in those low-lying areas (Figures 9.4a and b).

Drainage

In winter the remainder of precipitation (precipitation minus evaporation) and the total salt load owing to seepage and salt intrusion must be drained out of the inland waters. An increasing sea level rise will induce an increasing total discharge owing to both an increase in salt load and an enlargement of the drained area, as well as a growing difference between sea and polder level. This will entail extra costs upon drainage, depending on the increase in total discharge and the sea level rise. The relation between these two is supposed to be linear according to Abrahamse et al. (1982a,b, c), and de Jong (1986).

$$D_t(i) = (C_1 * \Delta U_t(i) + C_2 * Q) * (SLR_t + H) \tag{9.8}$$

with:
$D_t(i)$ = extra drainage costs for area i at time t (in Dfl/yr)
$i = 1, \ldots 7$ (seven drainage segments)
C_1 = constant of area i (in Dfl/m^3.m.yr)
C_2 = constant of area i (in Dfl/m^3.m.day)
$\Delta U_t(i)$ = increase in discharge in area i (in m^3)
Q = daily water amount in area i (in m^3)
SLR_t = sea level rise (m)
H = raising height in view of resistance (in m)

IJssel Lake

Presently, there is a natural water discharge from the IJssel lake to the Wadden Sea, caused by the head difference between the sea level of the IJssel lake and the Wadden Sea at low tide (−0.9 m N.A.L.). An accelerated sea level rise, exceeding 0.4 meter, will drastically disturb this natural discharge. Measures have to be taken in order to restore the water balance, of which three different options are proposed in den Elzen and Rotmans (1988), namely:

- drainage of the IJssel lake, with high drainage costs;
- raising of the level of the lake, requiring an equivalent corresponding raising of the dikes around the lake;
- digging a canal through the lake, necessitating a dike construction through the lake.

De Ronde and De Vrees (1990) argue that the drainage of the IJssel lake would be the best solution, because of the flexibility and the possibility of spreading the costs for changing the infrastructure. Within 30 years the situation of the IJssel lake might come to a head, and then a choice from the possible options must be made.

9.3.4 Agriculture

Temperature is one of the major growth factors in agriculture. Crops differ in their heat and water requirements. For some crops, like seed plants, a surplus of degree days caused by a temperature increase will decrease the yield, as a consequence of a shortened growth season, while for others the growing season will be extended. Nowadays in the Netherlands there is a precipitation shortage during summer, which will only become worse in the expected new hydrological conditions (Den Elzen and Rotmans, 1988).

The preliminary research of Goudriaan (1988) into the consequences of the greenhouse effect on agriculture for the Netherlands shows that there may be a complex shift in agricultural species, with predominantly positive effects. The greatest disadvantages arise from a possible prolonged dryness in summer. Positive aspects come from the stimulating effect of the increased carbon dioxide concentration on the photosynthesis of most crops. The production growth (cropped portion) of most $C3$ plants by increased CO_2 will amount to 0.17 percent per year per *ppm* CO_2 (Health Council of the Netherlands, 1986). However climate changes as well as other air pollutants

9.3. QUANTIFICATION OF IMPACTS FOR VARIOUS SECTORS

might undo this positive effect. Unambiguous climate/crop models have not yet been developed. In spite of this it can be tentatively concluded that the Dutch agricultural sector will not suffer greatly under the direct harmful consequences of climate changes. It will be rather influenced by new technological developments, and environmental measures adopted by the European Community.

9.3.5 Energy Use

Climate and energy are linked. On the one hand, climate changes influence energy use (heating, cooling and drainage); and on the other hand, climate is influenced by energy use (greenhouse effect). In Den Elzen and Rotmans (1988) the consequences of a temperature increase have been examined for the cooling and space heating sectors, the latter being broken down into heating for households and offices.

The energy saving for households by increased temperature is calculated with a simulation model developed by the Dutch Organization for Applied Scientific Research, TNO (Dubbeld, 1985), resulting in the following expression:

$$\Delta E(t) = \Delta T_{eq}(t) * (0.126 * ENSP + 31.73) \qquad (9.9)$$

with:
$\Delta E(t)$ = energy saving of a mean household in the Netherlands at time t (in m^3/yr)
$\Delta T_{eq}(t)$ = global mean equilibrium temperature increase at time t (in $°C$)
$ENSP$ = energy use for space heating for a mean household (about 1900 gas m^3/year in the Netherlands according to Huizinga, 1988) (in m^3/yr)

This linear relationship does not hold beyond a temperature increase of 5 $°C$. In the heating sector the temperature increase may lead to high profits in energy saving. Quantifying these profits is difficult, because the future fuel and gas prices are unknowns. Calculations presented here assume no change in energy use and insulation of houses. Besides, apart from a temperature increase, remaining future climate conditions are supposed to be unaltered. Future fuel and gas prices are based on a moderate (middle) scenario of the Dutch Central Planning Bureau (Huizinga, 1988). On the other hand the reduced demand for heating would create a loss for the Dutch economy. Therefore it is still unclear how this decrease in heating demand will affect the Dutch economy.

In the cooling sector we use a subdivision into cooling of food (refrigerators, freezers and cold stores) and space cooling (air conditioning) (Den Elzen and Rotmans, 1988). A temperature increase would mean additional costs for cooling and investment costs for air conditioning in public and private buildings (Gibbs et al., 1987). It is not yet possible to give reliable cost estimates for the cooling sector. In conclusion, weighing the costs and benefits for the energy use sector, the costs do not seem to outweigh the profits.

9.4 Results

As mentioned before the results are based on the four scenarios described in section 2.3. In the pictures only the range built up from the scenarios A and D is represented. Some results are shown by means of a map of the Netherlands, analogous to those given in the Figures 9.4a, 4b and 4c.

Coastal Defence

The frequency of overtopping the flood protection for the threatened area in the present situation as well as in a situation with a sea level rise of 0.5 meter, or 1.0 meter, respectively, is given in Figure 9.5. The frequency of overtopping in 2100 will increase from 0.0016 for scenario D (forced trend) to more than 0.021 for scenario A (unrestricted trend) for Central Holland; the latter value goes far beyond the statistical chance of inundation of the disastrous big storm surge in 1953. Figure 9.6 shows how the safety of the Western coastal zone (Central Holland) will change by application of the strategy of optimal dike raising for the scenarios A and D. The path of frequency is sawtoothed, because of the sudden drop after each dike raising. After this the frequency will slowly increase, under the influence of a continuing sea level rise. The frequency at the end of the simulation ranges from 0.0000127 (scenario D) to 0.000044 (scenario A), being averaged only a quarter of the present Delta norm of 0.0001. Because this 'average' frequency of about 0.000028 forms the result of an economic analysis, which takes into account the future worth of material goods and people at risk, this value can be considered as a new safety level. However, this value is much lower than the Delta norm. This means that, at the end of the next century, after having simulated several dike raising steps, the increased economic worth of the threatened area, necessitates a stricter safety norm. Based on the foregoing it can be concluded that the Delta norm should be tightened up.

9.4. RESULTS

Figure 9.7 shows the calculated step-wise dike raising over the period 1990 to 2100 for the Western zone (Central Holland) for scenarios A and D. The number of steps and the raising per step depends on the total optimal dike raising and the minimum adaptation of a dike, which has been chosen as 30 centimeters. The total dike raising for the period 1990 to 2100 for scenario A is about 2.40 meters (Central Holland), 2.30 meters (Zeeland), and 2.10 meters (Friesland and Groningen). For scenario D the dike raising in 2100 is about 1.80 meters (Central Holland), 1.70 meters (Zeeland), and 1.55 meters (Friesland and Groningen). The total costs involved amounts up to about 20 billion guilders.

In both cases this optimal dike raising is added with extra adaptations due to high tide level and higher waves, of about 55 cm. The relatively minor difference between the total dike raising in scenario A and D is caused by considering the dike raising issue as an economic decision problem.

Another strategy which could have been used is that of maintaining the safety level at the present "Delta norm" level (Den Elzen and Rotmans, 1988, and Den Elzen and Rotmans, 1990). Using this strategy gives a dike raising in the year 2100 of 1.55 m for scenario A, and 0.80 m for scenario D. Now the maximum difference between the total dike raising for the two different strategies is about 1.60 m. For this strategy the total costs vary from about 15 to 18 billion guilders.

Dune Strengthening

Figure 9.8 gives the costs of dune erosion for the period 1990 to 2100 for both scenarios. The total cost of dune strengthening will increase up to 5 billion guilders for scenario D to nearly 6.5 billion guilders for scenario A at the end of the simulation period (2100).

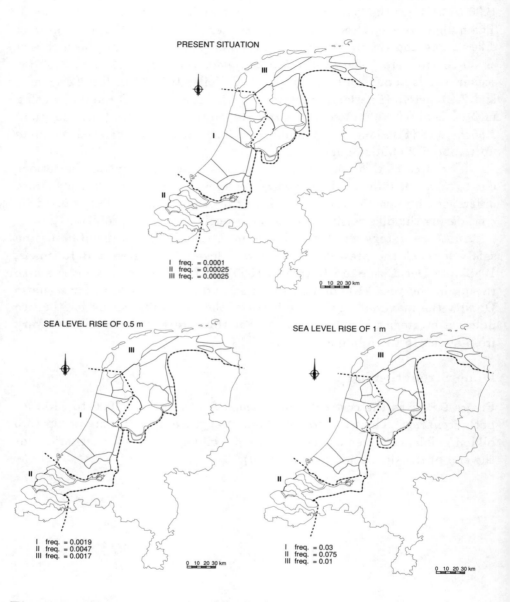

Figure 9.5: Flooding frequency (safety) of the threatened area in the present situation and in case of a sea level rise of 0.5 and 1.0 m.

9.4. RESULTS

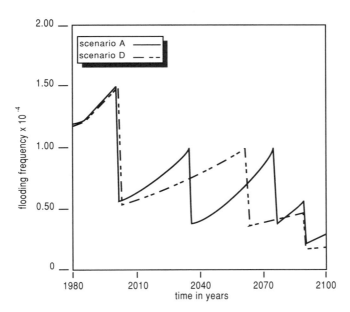

Figure 9.6: Flooding frequency for scenarios A and D for the Western coastal zone (Central Holland).

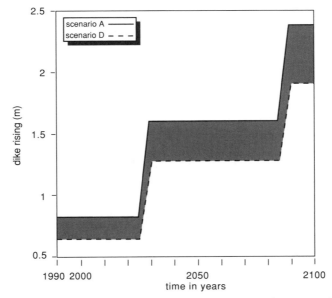

Figure 9.7: Simulated step-wise dike raising for scenarios A and D for the Western coastal zone (Central Holland).

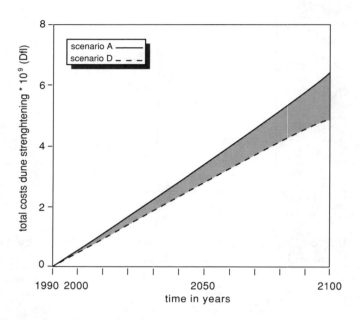

Figure 9.8: Simulated total costs of dune strengthening for scenario A and D.

Water Management

Figure 9.9 gives the change in length of the salt tongue, measured from the Hook of Holland for a low Rhine discharge (1000 m^3/s) for scenarios A and D. The Dutch IJssel is situated at 30 kilometers from the Hook of Holland and with the situation as it stands the Dutch IJssel is not being penetrated by salt water. If sea level rise were to continue, the Dutch IJssel would become brackish for both scenarios, as Figure 9.9 shows. At another low discharge rate ($> 1000 m^3/s$) the Dutch IJssel will be intruded only for scenario A. Generally, in case of a continuing sea level rise, the Dutch IJssel will be intruded more frequently, which would cause high agricultural damage.

Figure 9.10 illustrates both the present and the future seepage situation for the Netherlands at a sea level rise of 0.5 and 1.0 meter, respectively.

9.4. RESULTS

The total amount of seepage increases by 40% for scenario D and by 80% for scenario A. The total present salt damage caused by seepage and salt intrusion, about 533 million guilders, will be nearly doubled, as the sea level rise amounts to 1 meter.

In Figure 9.11 the calculated drainage costs are depicted. In case of a 1 meter sea level rise the present costs, 12 million guilders, will increase to about 55 million guilders.

In Table 9.2 the pros (profits) and cons (costs) of the different options for the IJssel lake are represented by means of a multi-criteria analysis. Based on such an analysis, the digging of a canal through the lake, turns out to be the best solution. However, such a construction would bring about a dramatic change in the lake's infrastructure. Therefore priority should be given to the second solution.

Table 9.3 summarizes indications for the total costs for different scenarios (A and D). Actually the costs may be higher, because the influence on the fishing, shipping, industry, health and recreation sectors has not been taken into account.

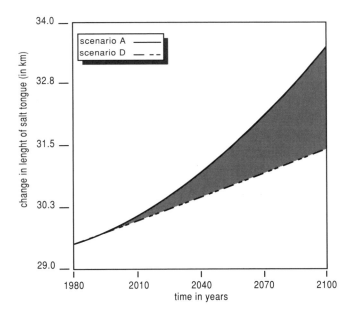

Figure 9.9: Simulated change in length of the salt tongue.

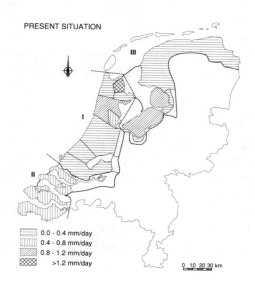
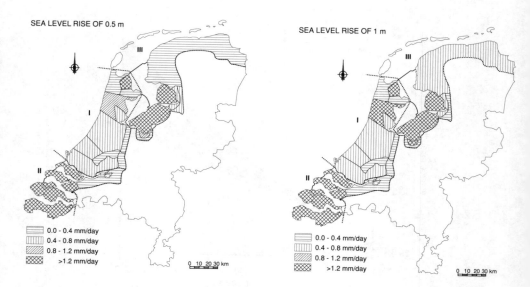

Figure 9.10: Present and future seepage situation in the Netherlands in case of a sea level rise of 0.5 m and 1.0 m.

9.4. RESULTS

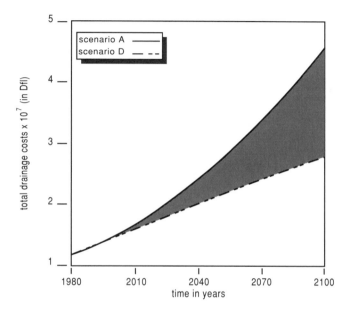

Figure 9.11: Simulated total drainage costs.

solution	safety	water supply	salt load	seepage	total costs
drainage of the IJssel lake, with the same level of the lake	raising the main dike, with the sea level rise $-$	unchanged 0	$(z/0.4) * 16425$ (ton Cl/year) $-$	increase of seepage along the main dike 0	1. raising the main dike $K = 30 * 10^6$ $*(1 + 5 * z)$ 2. drainage a. investment $2.79 * 10^9$ b. yearly $(27.9 + z * 14.7) * 10^6$
raising the level of the IJssel lake	raising dikes around the lake (200 km) dike and dikes along the IJssel and Vecht $--$	extra amount of water for water supply $z * 1.2 * 10^9 m^3$ $+$	unchanged 0	increase of seepage to areas around the IJssel lake $--$	1. raising the main dike $K = 30 * 10^6$ $*(1 + 5 * z)$ 2. drainage a. investment $1.6 * 10^9$ b. yearly $13.5 * 10^6$ 3. dike raising $K = 200 * 10^6$ $*(1 + 5 * z)$
construction of a canal through the IJssel lake	raising the dike Den Oever – Enkhuizen, the main dike and the dikes along the IJssel and Vecht $--$	extra amount of water for water supply $z * 0.5 * 10^9 m^3$ $+$	unchanged 0	increase of seepage to Medemblik Flevoland and Wieringer Lake $-$	1. raising the main dike $K = 30 * 10^6$ $*(1 + 5 * z)$ 2. drainage a. investment $1.6 * 10^9$ b. yearly $13.5 * 10^6$ 3. costs of construction of a dike through the lake

Table 9.2: Countermeasures for the IJssel lake with a sea level rise (of z meter)
$--$ = strong negative effect
$-$ = negative effect
0 = no effect
$+$ = positive effect
$++$ = strong positive effect

	total costs $*10^9$ for period from 1990 till 2100	
	scenario A	scenario D
dikes	20 Dfl.	15 Dfl.
dunes	6.5 Dfl.	5 Dfl.
intertidal area	++	+
salt intrusion	++	+
seepage	+80% increase in 2100	+40% increase in 2100
drainage	3 Dfl.	2.5 Dfl.
agriculture	+/−	+/−
energy	−	−
IJssel lake	4.5 Dfl. +	4 Dfl. +
	extra costs (++)	extra costs (+)
shipping	++	+

Table 9.3: Indicative costs for different scenarios (A and D) for several sectors of Dutch society
+   = increase in costs
++ = strong increase in costs
−   = decrease in costs

9.5 Conclusions

Simulations with IMAGE suggest that the socio-economic consequences of the greenhouse effect for the Netherlands are large, involving considerable cumulative costs for the next century. However, the yearly costs will be less than a half percent of the gross national product. This is in line with a recent study by Delft Hydraulics (1990).

Our calculations show that for all scenarios the dikes will have to be raised before 2000, otherwise the safety will decrease far below the Delta norm (the present safety level). Applying the strategy of optimal dike raising, which is preferable from an economic point of view, it follows that the minimum extra dike raising for the next century is 2 meters. This would imply high costs, which amount to about 20 billion guilders. A major inference is that the Delta norm is too low and should be tightened up to 0.000025, a quarter of the present value, entailing a better protection of human lives and material goods. This results in an extra dike raising of about 0.55 meter.

In order to avoid a further, socially unacceptable retreat of the Dutch coastline, a policy review to a policy that arrests dune erosion is recommended. In that case the present yearly costs of dune erosion, about 30 million Dutch guilders, will in the future increase to at least 350 million guilders and at most 600 million guilders. The total cost of coastal defence range, depending on which scenario and which strategy will be chosen, from a few billion up to 30 billion guilders for the period from 1990 to 2100.

For the water management sector the intrusion of salt water as well as the increasing seepage will become major future problems, especially for the agricultural sector. Additionally the yearly drainage costs will increase considerably and the situation for the IJssel lake is getting worse. Having regard to the costs and the infrastructure of the lake the best solution is to raise the level of the IJssel lake. According to model calculations with IMAGE the total extra costs that have to be spent on the water management due to a sea level rise are expected to be more than ten billion guilders.

Although not pretending to give a comprehensive overview of the consequences of the greenhouse phenomenon for the Netherlands, this study gives indicative figures concerning the socio-economic consequences. In general, for the Netherlands the socio-economic costs induced by the greenhouse effect will be considerable, but can be kept under control if Dutch policy makers will anticipate in time.

Chapter 10

Policy Analysis

10.1 Introduction

Even faster than the depletion of the ozone layer the anticipated climate change has come to play a role in high level policy debates. Since the problem is very complex by nature a need has arisen among policy makers to have at their disposal a tool that gives a clear and concise overview of the workings of the greenhouse effect and the relevance of potential policy options. The Integrated Model to Assess the Greenhouse Effect (IMAGE) was developed at the National Institute for Public Health and Environmental Protection (RIVM in Dutch) from 1986, at a stage when the recognition of climate change was intended to be its primary role (Rotmans, 1989). Since the enhanced greenhouse effect is created by a multitude of effects, the model tries to capture these causes and effects in an integrated fashion. The main difficulty thwarting policy response to the greenhouse effect is that the causes are not only many; they also form the fundamental basis of our society: the present practices with respect to energy production and consumption, agriculture and industry. Dependent on the assumptions and definitions it can be generally said that the agricultural sector, including deforestation, causes 25% of the problem, while 75% is caused by the energy and industry sectors. Within the energy sector transportation, power generation and other combustion processes play about an equal role (see Figure 10.1).

In the last few years IMAGE has been used to evaluate long-term climate strategies. In this chapter some calculation examples, made for the Intergovernmental Panel on Climate Change (IPCC), are presented.

194   CHAPTER 10.   POLICY ANALYSIS

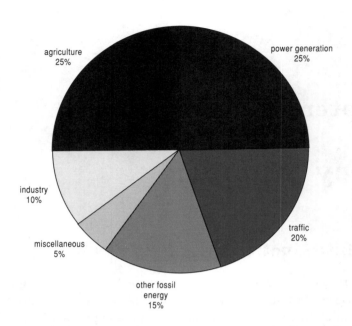

Figure 10.1: Relative contribution to climate change.

10.2   Scenario Calculations

The main application of IMAGE is the development and evaluation of long-term scenarios, up to 2100. It has been used extensively in the Netherlands over the last few years to evaluate different future world options with respect to climate change. Besides demonstration sessions for different groups the model was used for different Dutch environmental studies, such as the preparatory document for the National Environmental Plan (Langeweg, 1989).

Recently, IMAGE entered the international arena. Some scenario results from calculations made within the framework of the Intergovernmental Panel on Climate Change (IPCC) will be briefly discussed here. In early 1989 the Steering Committee of the Response Strategies Working Group of IPCC requested the USA and the Netherlands to prepare a document on three emissions scenarios for use in different IPCC working groups. Initially these scenarios had to be designed in such a way that they would lead to a doubling of CO_2 equivalent concentrations in the years 2030, 2060, and subsequently in 2090. In the last scenario the concentration would stabilize afterwards. The Dutch contribution to the draft report (IPCC, 1989), which was prepared jointly with EPA, consisted, besides the determination of the conceptual approach, of the performance of a number of IMAGE calculations. Preliminary scenarios made with EPAs Atmospheric Stabilization Framework (EPA, 1989) were evaluated and, secondly, a number of additional analyses were made. Since the consequences of policies in the decades to come are of crucial importance for the purpose of IPCC, some runs were performed simulating delayed response actions.

10.3   Low Climate Risk Scenario

The Netherlands and some other countries were not satisfied with the range of scenarios selected, primarily since even the lowest scenario would lead to temperature increases that are unprecedented in the last 100,000 years and more. Not considering the feasibility of pursuing such a scenario (which is low) some calculations were made to access tolerable emissions which would limit climate change risks sufficiently. Such a low scenario is important for the assessment of the reaction of the climate system by scientists and is the first step towards painting a picture of a sustainable world. It is not yet included in the draft report referred to above.

At a workshop in Bellagio (Jäger, 1988) it was suggested to take a global rate of temperature of $0.1°C$ per decade as an initial target value, which would allow for adaptation of ecosystems. The German Physical and Meterological Societies recommended an abolute temperature change of 1 to 2 degree celsius temperature rise from pre-industrial levels (German Enquete Commission, 1988). These values are consistent with estimates of maximum natural climate variations in the past according to Sassin et al. (1988), who evaluated past rates of temperatures and abolute changes from ice core temperature records. Since no past evidence appears to be available for faster temperature changes than approximately $0.1°C$ per decade, we used this as a reference value for a policy of risk minimization. The resulting 'low climate risk' scenario can be considered as a first approximation of a long-term environmental goal enabling a sustainable development of the world economy.

The global low climate risk scenario induces a gradual decrease of global CO_2 emissions, which is more or less consistent with the Toronto recommendations (Environment Canada, 1988), if applied worldwide: a 20% reduction in 2005 and a 50% reduction by half way through the next century. Kram and Okken (1989) calculated that such reductions are possible in the Netherlands after 2020.

10.4 Results

In order to determine a CO_2 emission pathway complying with the above-mentioned rate of temperature change we assumed maximum control for the emissions of the other trace gases (Table 10.1), as estimated in the IPCC scenario leading to stabilization at double pre-industrial CO_2 equivalent concentration after 2090 (570 *ppm*).

	CO_2 (Gt)	CH_4 (Tg)	N_2O (Tg)	CFC-11 (Mkg)	CFC-12 (Mkg)	CO (Tg)
1985	5.40	510	11.3	296	433	2429
2000	4.32	538	11.8	217	280	2479
2025	2.70	579	12.0	32	22	2724
2050	2.70	584	12.0	21	21	2909
2075	2.70	540	11.7	22	23	3030
2100	2.70	471	10.2	24	26	3098

Table 10.1: Emissions of trace gases for the Low Climate Risk Scenario.

In Figure 10.2 the resulting low climate risk scenario is shown, together with the IPCC scenarios. It appears that the highest IPCC scenario would allow an increase of CO_2 emissions by a factor of 4 in 2100, while, in order to follow the lowest pathway, these emissions can only grow by about 40% towards the mid-21st century as compared to 1985. For comparison we drafted the expected carbon dioxide emissions associated with the most recent IEA World Energy Outlook (IEA, 1989). If world energy use and fuel mix were to follow this projected path, the emissions of CO_2 and other trace gases would exceed even the highest scenario as taken into account by the IPCC. To put the scenarios into an even more pessimistic perspective: in the lower scenarios it has been assumed that the emissions of the other trace gases can be limited within the same time schedule as CO_2, albeit at different levels. For regulated CFCs this might not cause problems, but for gases with as yet not fully quantified sources, like methane, nitrous oxide and carbon monoxide, this tends to be an optimistic assumption. If these limitations could not

10.4. RESULTS

be achieved, the necessary control of carbon dioxide should even be stricter than envisaged in the present IPCC scenarios in order to reach the same goals. In terms of geographical distribution the lowest scenario allows for increasing emissions in the developing countries and more or less stabilizing emissions in the industrialized countries. In the Netherlands National Environment Plan this stabilization (to be achieved in 2000 at 1989 levels) has been included as a provisional goal for the emissions of carbon dioxide.

The global low risk scenario could only be achieved by a massive and rapid change of the world energy system towards maximum energy efficiency and the application of CO_2 free technology. Although such a change might lead to continuing or even accelerated economic growth for some groups in some countries, there is a risk in major parts of the world that these emission reductions might only be achieved by limitations to quantitative economic growth which are unacceptable from the viewpoint of the present day national ambitions.

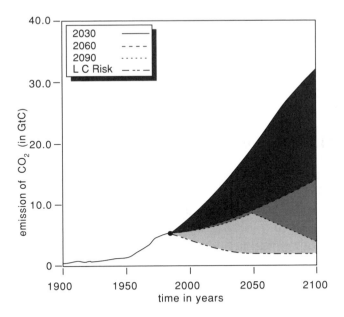

Figure 10.2: IPCC and Low Climate Risk Scenario.

A 20% CO_2 reduction in 2000 and a further reduction to 50% results in the concentrations for the different trace gases shown in Table 10.2.

	$CO_2 - eq$ (ppm)	CO_2 (ppm)	CH_4 (ppm)	N_2O (ppb)	CFC-11 (ppb)	CFC-12 (ppb)
1985	380	345	1.68	303	0.22	0.38
2000	405	355	1.85	312	0.37	0.62
2025	422	362	2.09	326	0.33	0.60
2050	429	365	2.29	339	0.26	0.52
2075	431	368	2.22	349	0.21	0.45
2100	421	371	1.76	353	0.17	0.40

Table 10.2: Concentrations of trace gases for the Low Climate Risk Scenario.

In Figure 10.3 the CO_2 equivalent concentrations as agreed upon by the IPCC Working Group are shown.

Figure 10.4 shows the resulting temperature curves for climate sensitivities of 2 and $3°C$. These results are thus consistent with the Toronto recommendations provided that the reductions are implemented worldwide and also measures are taken to control the emissions of the other greenhouse gases. The IPCC 2090 stabilization scenario would allow for slightly increasing emissions of CO_2, but would lead to temperature rises which are far above the suggested $0.1°C$ per decade.

10.5 Delayed Response

To underscore the importance of rapid decision making we performed a final analysis, in which we delayed the start of international response actions from the present to 2000, 2010, 2020 and 2030, respectively. It has been (optimistically) assumed that greenhouse gas emissions would follow the emissions of the 2030 doubling case should no action be taken to limit greenhouse gas emissions. Furthermore it is assumed that if the international decision is taken and followed up to start controlling climate change, the policy target would be stabilization of concentrations in 2090 at double CO_2 equivalent levels (the lowest IPCC scenario).

10.5. DELAYED RESPONSE

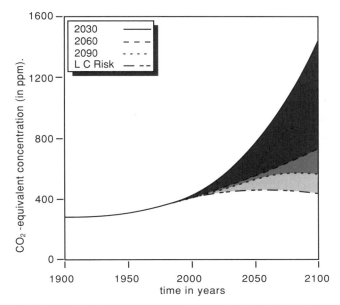

Figure 10.3: CO_2 equivalent concentrations for the IPCC scenarios and the Low Climate Risk Scenario.

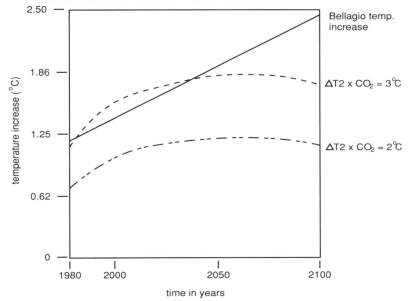

Figure 10.4: Equivalent temperature curves for the Low Climate Risk Scenario (for climate sensitivities of 2 and $3°C$) compared with the Bellagio temperature increase curve (i.e. $0.1°C$ per decade).

Emissions of non-CO_2 trace gases were assumed to be deflected towards the values associated with the 2090 scenario and consequently the allowable emissions of CO_2 were determined. In Figure 10.5 the emissions of CO_2 for different delayed response runs are depicted.

It appears that waiting until 2000 would imply only a slight necessary decrease of emissions. Even a slight increase of the emissions is permitted to reach stabilization, because a continuing decrease of the emissions should result in decreasing concentrations instead of stabilizing concentrations. In consequence of this 'stabilizing concentrations' target, the same temporary increase in emissions occur in the 2020 and 2030 responses. Waiting 10 years more would necessitate gradual emission reductions of about 40%. Again, 10 years later, in 2020, almost 70% emission reduction would have to be achieved in less than 10 years in order to reach the target. As can be expected, because of the choice of baseline emissions (leading to 2030 doubling), delaying response until 2030 would render a complete and swift phase-out of CO_2 emissions from fossil fuel combustion necessary. Should the $0.1°C$ per decade scenario be chosen as a target rate the emission reductions should even be more dramatic.

The resulting changes in atmospheric composition for the delayed response scenarios, as calculated by IMAGE, are represented in Figure 10.6. Only for the 2030 emission response scenario is it nearly impossible to stay under the target CO_2 equivalent scenario of 570 *ppm*.

10.6 Future Worlds

What kind of world would meet the requirements of the 'tolerable' scenario presented above? Although an infinite number of worlds with different combinations of energy systems and agricultural practices with associated emissions of greenhouse gases would meet the target, some brief indications will be given of how the emission scenarios that fulfilled the requirements can be arrived at. World population is assumed to approach 10.8 billion in 2100. The Low Climate Risk Scenario can be simulated with the Edmonds and Reilly model in a number of ways. When assuming a per capita economic growth rate of 1.7% in the developed world and 2.9% in the developing countries the most important policies would include: an annual end use efficiency increase in the industrialized world of 1.5% and 1.8% in the third world, while at the same time stimulating solar energy (reducing the costs within 40 years to US$ 10.-/GJ), preventing production of synfuels by high

10.6. FUTURE WORLDS

non-energy costs and gradually introducing environmental taxes on end use proportional to the carbon content of the fuel type. Compared to the present situation, coal is fully phased out, oil supply is more or less stable, the use of natural gas increases, but the major part of the growth of the energy demand (moderated by large efficiency increases) is captured by renewables. To illustrate the dependence of these results of economic growth: when for instance Asian and Latin-American countries are assumed to achieve higher per capita economic growth (3.5% annually), efficiency increases in those countries should be higher (2.2% per year) and renewables introduced faster (solar costs US$ 8.-/GJ in 25 years) than in the previous analysis. The final implication of a CO_2 emission reduction strategy would be a shift away from fossil fuels, implying that the available resources will not be fully used. It is a major question if the global energy system, which is now based for more than 80% on fossil fuel consumption will be able to cope with the social and economic effects of such a shift.

For CFCs we assumed an almost total phase-out with the exception of a number of developing countries not participating in the Montreal Protocol. To arrive at the assumed more than 95% production decreases the Protocol would have to be strengthened. Methane emissions are assumed to be stabilized by limiting the consumption of meat and dairy and gradually (applying logistic curves) increasing recovery of methane losses from coal mining and waste dumps to 50%. Anthropogenic CO emissions are limited by technological measures for vehicle exhausts and in industry. Growth of anthropogenic N_2O emissions from fertilizer application is controlled by balancing increases in fertilizer consumption in the developing world against decreases in the developed world enabled by modifications in agricultural practices.

CHAPTER 10. POLICY ANALYSIS

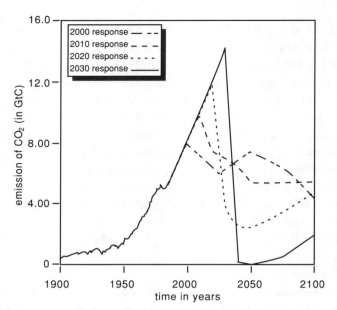

Figure 10.5: CO_2 emissions for delayed response analysis.

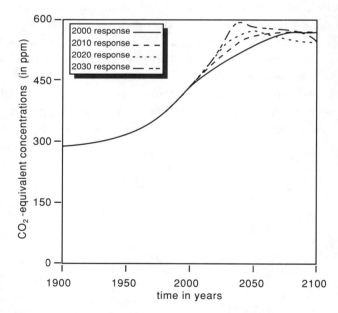

Figure 10.6: CO_2 equivalent concentrations for delayed response analysis.

10.7 Conclusions

Calculations with IMAGE suggest that the rate of global temperature change can be limited to values which do not go beyond past climate experience by the global implementation of CO_2 emission reductions of 20% in 2000 and 50% in 2025. These figures are consistent with the recommendations of the Toronto Conference on the Changing Atmosphere in 1988, provided these reductions are applied globally and a strong effort is undertaken to control the emissions of the other greenhouse gases, including a near phase-out of CFCs.

The projections of IEA (IEA, 1989) show that if the greenhouse effect is not taken into account in energy policy planning, emissions of greenhouse gases will show accelerated growth if the present world economic growth continues, based on the same energy system as before. Abundant reserves of cheap coal will be used unless technological and financial mechanisms are developed to push energy supply development in a more benevolent direction.

According to the IMAGE calculations the different lag effects in physical and socio-economic processes make a very rapid initiation of international policies necessary in order to limit major risks of climate change. Although the necessary changes appear to be technically feasible, major social and economic impediments will have to be removed to achieve the recommended reductions, but a complete phase-out of fossil fuels would not be necessary. IMAGE simulations show that the emissions in the industrialized countries should be decreased by at least 80% over the coming 50 years in order to prevent serious risks to the world community. Delays in policy formulation will be followed by long lead times for introduction of measures, increasing the amount of climate change 'in the bank' because of the slow response of the physical climate system. Therefore rapid response action is warranted.

In order to evaluate recent and future proposals for control of greenhouse gas emissions it is recommended to determine a long-term environmental or climate goal that would allow for a sustainable development of the world economy.

Chapter 11

Temperature Increasing Potential

11.1 Introduction

In order to develop environmental long-term goals with respect to climate change an index is needed to compare the temperature increasing effect of greenhouse gas emissions. Here the concept of Temperature Increasing Potential (TIP) is introduced as a greenhouse counterpart to the ozone depleting potential (ODP). To obtain the relationship between an emission and its associated effect on temperature both the model approach (IMAGE) and the analytical approach is used. Furthermore, both approaches are compared to previous efforts to determine relative greenhouse gas potentials.

11.2 Relation between Temperature and Emissions

In deriving emission targets from a set goal for global mean temperature increase, many nonlinear relationships within the atmosphere have to be considered. The emission of greenhouse gases initially leads to increased atmospheric concentrations. These gases are removed by a diversity of processes, varying with each gas and its atmospheric concentration: uptake by oceans, deposition, photochemical reactions, uptake by biota and soils. These removal processes determine the atmospheric lifetime of the gases. Furthermore, many other factors related to the greenhouse problem interact with the removal processes; for example, the concentration of other energy-related gases like carbon monoxide (CO) and non-methane hydrocarbons and the influence of climate change on the carbon cycle and on methane (CH_4) release from natural reservoirs. Additionally the radiative absorption rate is neither constant nor proportional to their respective concentrations. Finally these processes and their underlying assumptions are scenario dependent. Generally, in order to compare the result with previous efforts (e.g. Lashof and Ahuja, 1990, Derwent, R.G., 1990, Fisher et al., 1990) equilibrium temperature effect will be used instead of transient responses.

The relation between an emission and its associated effect on temperature can be expressed in terms of Temperature Increasing Potential (TIP), or Global Warming Potential (GWP), as used in the literature. This is comparable to the ozone depletion potential (ODP) defined by Wuebbles (1981) which interrelates different ozone-depleting substances. However the actual TIP is time-dependent, and not a scalar constant with which one could multiply emissions like the ODPs. Nevertheless, for want of a better alternative, the relative radiative potential of the trace gases will be approximated by a scalar.

11.3 Methodology

11.3.1 Definition

To achieve a direct relationship between an emission of a greenhouse gas and its corresponding temperature response the following strategy will be used. For the greenhouse gases CO_2, CH_4, and N_2O, one emission impulse of 1 Gt will be generated during one year, the year 1986. For CFCs emissions a pulse of 10^{-3} Gt will be posited during one year, because the emissions of CFCs are of magnitude 10^3 till 10^4 lower than of the other gases, and a pulse of 1 Gt might lead to systems instabilities. Of course the corresponding pulse of CO_2 will in case of CFCs also be equal to 10^{-3} Gt. To test the sensitivity of this modelling approach to variations in the emission pulses, simulation runs with variations in both magnitude and time-span have been carried out. It appeared that the method is fairly robust for variations in magnitude and time-span of emission pulses (Rotmans and Den Elzen, 1990). Grams and not moles are used, because in the international literature emissions are mostly expressed in grams.

The Temperature Increasing Potential, or TIP, of a greenhouse gas is then defined as the integrated temperature effect (which consists of the integral of time-dependent temperature distributions from 0 to a time t) of 1 Gt

11.3. METHODOLOGY

emission of that specific gas compared to that of CO_2:

$$TIP_i(t) = \frac{\text{integr. temp. effect of 1(or } 10^{-3}) \text{ } Gt \text{ of trace gas } i \text{ at time } t}{\text{integr. temp. effect of 1(or } 10^{-3} \text{ } Gt \text{ of carbon dioxide at time } t} \quad (11.1)$$

with:
$TIP_i(t) = $ temperature increasing potential of trace gas i at time t

Other possible definitions are given in Lashof and Rotmans (1990).

In determining the TIP, two quintessential matters must be considered. First the influence of the rather arbitrarily chosen time-span and height of the emission impulse. To measure the influence of various kinds of pulses a sensitivity analysis has been carried out with emission impulses of 0.25, 0.50, and 1.0 Gt (and 0.25 10^{-3}, 0.5 10^{-3}, and 10^{-3} for CFCs), during 1, 5, and 10 years, respectively. The results of this analysis are presented in Rotmans and Den Elzen (1990). Secondly, the target point in time of the TIP, being a crucial aspect in the TIP analysis, has to be determined. The time dependency of the TIP is mainly due to the fact that CO_2 does not have a specific atmospheric lifetime, but is exchanged between atmosphere, ocean, and terrestrial biosphere. To overcome this problem two case studies will be treated, one with a relatively short time horizon (instead of atmospheric residence time) with respect to CO_2 of 100 years, and a second one with an extremely long time horizon of CO_2 of 500 years. So only the limits of the integration are varied, not the internal dynamics of the CO_2 model.

According to these assumptions and based on definition (11.1), the TIP can be calculated in two different ways. Earlier attempts were based on simple analytical approaches (Lashof and Ahuja, 1990), which directly calculated the temperature effect from the emissions. Here also an indirect analytical method is presented, calculating concentrations first and then global temperature effects.

An alternative way of solving the TIP problem, which has not been applied before, is using integrated greenhouse models, relating emissions to global temperature rise. Presently there are three such integrated greenhouse models: IMAGE (Integrated Model to Assess the Greenhouse Effect, Rotmans et al., 1990a), the Model of Warming Commitment of the World Resource Institute (Mintzer, 1987) and the Atmospheric Stabilization Framework of the U.S. Environmental Protection Agency (EPA, 1989), which have recently been compared (AGGG, 1990). The models produced very similar temperature results for the same emission inputs for different trace gases, although very different approaches have been chosen for the representation of the carbon cycle, atmospheric chemistry processes, and other model aspects (Response Strategies Working Group, 1989a and 1989b). Therefore, notwithstanding the fact that these models embrace many uncertainties, international consensus on assumptions and methodologies should be possible, based on the best available knowledge (Swart et al., 1989).

Here IMAGE is used to calculate the TIP concept. Both the analytical and modelling approach will be compared and evaluated.

11.3.2 Modelling Approach

IMAGE is used in order to determine the relative radiative potential of the greenhouse gases CO_2, CH_4, N_2O, CFC-11 and CFC-12. To reduce the influence of the type of scenario used, stabilization scenarios have been developed for all trace gases; these emission stabilization scenarios result in steady-state concentrations in the second half of the next century. In each case two stabilization scenarios are compared in pairs, one with an emission impulse of a certain trace gas, and one without such an emission impulse. An emission impulse induces an emission of 1 (or 10^{-3} Gt during one year (1986), and is zero both after and before 1986. An example of such a pair of emission stabilization scenarios is given in Figure 11.1, where a CO_2 emission scenario with and without impulse is depicted. Figure 11.2 gives the concentrations in pairs for CH_4.

Then two equilibrium temperature responses are simulated, again with and without an emission impulse. By subtracting these two temperature responses, the influence of the scenario choice is reduced, yielding the net temperature effect. This net temperature effect is integrated from time 0 (in 1900) to the time horizon. Dividing the integrated net temperature effect of CH_4, N_2O or CFCs by that of CO_2, gives the TIP.

Specifically to simulate the TIP-concept, the usual simulation time-span, covering 200 years, from 1900 to 2100, is extended to the year 2500. The year 2500 relates to the chosen "endless" CO_2 atmospheric residence time of 500 years. The various assumptions for the different trace gases are given in Table 11.1 :

11.3. METHODOLOGY

	Concentration in 1985	Residence Time	Instantaneous Forcing	Conversion Factor
CO_2	346 (ppm)	–	0.0107 (°C/ppm)	0.471 (ppm.Gt)
CH_4	1.70 (ppm)	time-dependent	0.333 (°C/ppm)	0.376 (ppm/Gt)
N_2O	307 (ppb)	170 (yrs)	1.83 (°C/ppm)	0.200 (ppm/Gt)
CFC-11	0.22 (ppb)	75 (yrs)	190 (°C/ppm)	0.046 (ppb/Tg)
CFC-12	0.38 (ppb)	125 (yrs)	220 (°C/ppm)	0.048 (ppb/Tg)

Table 11.1: Survey of parameter values in IMAGE.

The radiative forcing factors are instantaneous radiative perturbations and are based on Ramanathan (1985) and Wigley (1987), and are slightly different from those given in Rotmans (1986). These values are valid for the ranges 280–390 *ppm* for CO_2, 0.90–1.70 for CH_4, 285–500 *ppb* for N_2O, 0–30 *ppb* for CFC-11, and 0–40 *ppb* for CFC-12. These radiative perturbations implicate a climate feedback sensitivity of 1.44 W/m^2 °C.

For CO_2 the atmospheric stabilization scenario is fed into the integrated carbon cycle module of IMAGE, consisting of the coupled ocean, terrestrial biota, and deforestation module. The stabilization scenario includes a sharp decrease of fossil fuels as well as a moderate deforestation scenario.

In the CH_4-CO-OH cycle module of IMAGE the concentrations of OH radicals and CO are maintained at a constant 1985 level. A substantial fraction of the increase in the methane concentration in the atmosphere is most probably caused by CO competing for OH radicals (Rotmans and Eggink, 1988). To measure this influence of CO on CH_4, an emission impulse of CO, 1 Gt in 1986 only, is also generated. Because methane emissions have an increasing effect on tropospheric ozone and stratospheric water vapor concentrations, the temperature effect of CH_4 is assumed to be enhanced by 70% (Lashof and Ahuja, 1990), represented by a factor 1.7. The temperature effects of CH_4 with and without a CO emission impulse are then compared to each other (Rotmans and Den Elzen, 1990).

The CFCs modules in IMAGE include a delay time between production and emission, which is assumed to be different for different applications. The extra impulse is added to the emission and not to the production, and so has no time delay.

Finally N_2O concentrations are computed from the emissions by taking into account an exponentially delayed emission mechanism, and a constant atmospheric lifetime of 170 years.

210   CHAPTER 11. TEMPERATURE INCREASING POTENTIAL

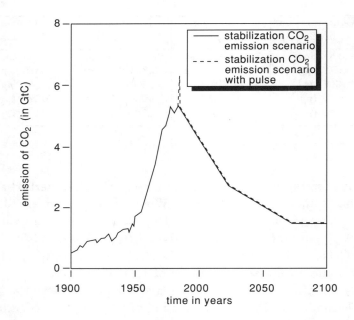

Figure 11.1: CO_2 emissions scenario with and without impulse.

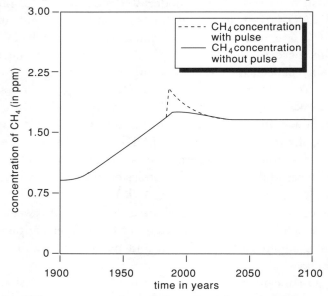

Figure 11.2: CH_4 concentration with and without impulse.

11.3. METHODOLOGY

11.3.3 Analytical Approach

Next to the modelling approach two different analytical methods are introduced, mainly to verify the results computed with IMAGE. To determine straightforwardly the relationship between the emission and the temperature effect of a trace gas, definition (11.1) is transformed into an analytically useful form:

$$TIP_i(t) = \frac{CVF_i * TMP_i * \int_0^T [EM_i(t) - RMV_i(t)]dt}{CVF_c * TMP_c * \int_0^T [EM_c(t) - RMV_c(t)]dt} \quad (11.2)$$

with:

$TIP_i(t)$ = temperature increasing potential of trace gas i at time t, whereas $t = 0$ represents the year 1985
CVF_i = conversion factor of trace gas i (in ppm/Gt)
TMP_i = radiative forcing factor of trace gas i (in $°C/ppm$)
T = time horizon (in years)
$EM_i(t)$ = global emission of trace gas i at time t; is here an emission of 1 (or 10^{-3}) Gt in 1986 and 0 elsewhere (in Gt/yr)
$RMV_i(t)$ = atmospheric removal of trace gas i at time t (in Gt/yr)

CVF_c, TMP_c, $L(c)$, EM_c, RMV_c are the corresponding values for CO_2 in the denominator.

The parameter values of CVF_i, CVF_c, TMP_i and TMP_c are identical to those used in the modelling approach and which are represented in Table 11.1. The global emissions, denoted by EM_i and EM_c respectively, are emission pulses: 1 (or 10^{-3}) Gt in 1986, and zero before and after this year.

The atmospheric removal process of CO_2 is reflected by the airborne fraction, defined as the fraction of the CO_2 emission that remains in the atmosphere, which is simply assumed to be a constant, 60%. For N_2O and CFCs the atmospheric retention can be described by a single atmospheric residence time. The atmospheric removal is supposed to be proportional to the concentration of these trace gases. This means that the fraction of a trace gas i that remains in the atmosphere can be represented by $e^{-t/L(i)}$. For CH_4, however, this is a far from realistic representation, in light of the complex atmospheric-chemical interactions with CO and OH. Therefore, instead of an exponential mechanism, a delayed exponential relationship is introduced, represented by $e^{-(t-\delta)/(L(i)+n)}$, where δ represents the time delay (5 years), and n represents the lengthening of the atmospheric residence time (1 year). These values of δ and n are calibrated on simulations with the CH_4 module of IMAGE. Lashof and Ahuja (1990) obviate this inadequacy by assuming that each mole of CO emitted increases the atmospheric CH_4 concentration by 0.09 moles. Following Lashof and Rotmans (1990), the temperature effect of methane is enhanced by 70%, due to the forming of tropospheric ozone and stratospheric water vapor.

The second way of analytically calculating the TIP includes, starting from emissions, a simplified calculation of the greenhouse gas concentrations and the resulting equilibrium temperature rises. This is similar to the modelling approach, and can be considered as an analytical approximation of the simulation method. This analytical approach has the advantage that the temperature effect, derived from emissions, is independent of any emission scenario. The total temperature effect over the atmospheric lifetime of each gas is calculated, both for an emission impulse and without an emission impulse. This emission impulse is 1 (or 10^{-3}) Gt for each trace gas in the year 1985. It should be noticed that metamodelling (Rotmans and Vrieze, 1990) is an outstanding method for determining a relationship between a greenhouse gas emission and its induced temperature effect. It is intended to work out this concept in the near future.

This analytical method, being the analytical equivalent to the modelling approach, involves another analytical interpretation of the TIP definition in (11.1):

$$TIP_i(t) = \frac{\int_0^T [\Delta T_i^{im}(t) - \Delta T_i^{wim}(t)] dt}{\int_0^T \Delta T_{CO_2}^{im}(t)} \quad (11.3)$$

with:

$TIP_i(t)$ = temperature increasing potential of trace gas i at time t

$\Delta T_i^{im}(t)$ = equilibrium temperature change with an emission impulse of 1 (or 10^{-3}) Gt of trace gas i, at time t ($^\circ C$)

$\Delta T_i^{wim}(t)$ = equilibrium temperature change without an emission impulse, at time t ($^\circ C$)

$\Delta T_{CO_2}^{im}(t)$ = equilibrium temperature change with an emission impulse of 1 (or 10^{-3}) Gt CO_2, at time t. N.B.: the equilibrium temperature effect of CO_2 without an emission impulse is zero, contrary to that of other gases, due to the logarithmical approximation of the radiative perturbations for CO_2.

T = time horizon

11.3. METHODOLOGY

CO_2

In a simplified manner the atmospheric concentration of CO_2 can be approximated by the following equation, based on Equation (2.2):

$$pCO_2(t) = pCO_2(t-1) + \int_{t-1}^{t} [ATMCF * AF * FSEM(\tau)d\tau] \quad (11.4)$$

with:
$pCO_2(t)$ = atmospheric CO_2 concentration (in ppm)
$ATMCF$ = factor that converts emissions of CO_2 into concentrations; is 0.471 ppm/GtC according to Brewer (1983) (in ppm/GtC)
$FSEM(t)$ = fossil fuel combustion flux at time t (in GtC/yr)
AF = airborne fraction, assumed to be constant, 0.60

Then the total equilibrium temperature effect integrated over the time horizon of CO_2 can be expressed as (see Appendix):

$$T_{CO_2} = T_{2xCO_2}/Ln(2) * T * Ln(1 + (ATMCF * AF * FSEM)/pCO_2(0)) \quad (11.5)$$

with:
T_{CO_2} = equilibrium temperature increase integrated over the time horizon of CO_2 (in $°C$)
T_{2xCO_2} = temperature increase for a doubled CO_2 concentration (in $°C$)
T = time horizon (in years)
$FSEM$ = emission impulse of 1 (or 10^{-3}) Gt CO_2 (in Gt)

CFCs

The atmospheric removal process of CFCs is calculated by a negative exponential function, with removal rate inversely proportional to the atmospheric lifetime of CFCs. Thus the atmospheric CFC concentration can be represented by the following expression:

$$pCFC(t) = e^{-t/LFTCFC} * (pCFC(0) + CVCFC * EMCFC) \quad (11.6)$$

with:
$pCFC(t)$ = atmospheric CFC concentration (in ppb)

$pCFC(0)$ = initial atmospheric CFC concentration, in 1985 (in *ppb*)
$LFTCFC$= atmospheric lifetime of CFC (in years)
$CVCFC$ = conversion factor of CFC (in *ppb/Gt*)
$EMCFC$ = emission impulse of 10^{-3} *Gt* CFC (in *Gt*)

Then the total equilibrium temperature effect can be derived from the difference in concentrations with and without an emission impulse of 10^{-3} *Gt* respectively (see the Appendix to this Chapter):

$$TCFC = ACFC * [(1 - e^{-T/LFTCFC}) * LFTCFC * CVCFC * EMCFC] \tag{11.7}$$

with:
$TCFC$ = equilibrium temperature effect of CFC integrated over atmospheric lifetime of CFC (in °C)
$ACFC$ = CFC temperature coefficient, obtained from Ramanathan et al. (1985) (in °C/*ppb*)
T = time horizon

N_2O and CH_4

In analogy to the atmospheric removal process of CFCs the removal of N_2O and CH_4 is described by a single residence time. For N_2O and CH_4 similar radiative perturbations are given by Wigley (1987), see the Appendix to this Chapter. This leads to the total equilibrium temperature effect:

$$\begin{aligned}TN_2O &= AN_2O * [2 * LFTN_2O * (1 - e^{-T/2*LFTN_2O}) * \\ &\quad (\sqrt{(CVFN_2O * EMN_2O + PN_2O(0))} \\ &\quad - \sqrt{(pN_2O(0))})\end{aligned} \tag{11.8}$$

with:
TN_2O = equilibrium temperature effect of N_2O integrated over atmospheric lifetime of N_2O (in °C)
AN_2O = N_2O temperature coefficient, obtained from Ramanathan et al. (1985) (in °C/*ppb*)
$LFTN_2O$ = atmospheric lifetime of N_2O (in years)
$CVFN_2O$= conversion factor of N_2O (in *ppb/Gt*)
EMN_2O = emission impulse of 1 *Gt* N_2O (in *Gt*)
$pN_2O(0)$ = initial atmospheric concentration of N_2O, in 1985 (in *ppb*)

And for CH_4:

$$TCH_4 = ACH_4 * e^{(\delta/(LFTCH_4+n))} * [2*(LFTCH_4+n) * \\ (1 - e^{-T/2*(LFTCH_4+n)}) * \\ (\sqrt{(CVF\ CH_4 * EMCH_4 + P\ CH_4(0))} - \sqrt{(pCH_4(0))}) \quad (11.9)$$

where $TCH_4, ACH_4, LFTCH_4, CVFCH_4, EMCH_4, pCH_4(0)$ correspond to the symbolic names of N_2O in Equation (11.8).

11.4 Results

The temperature increasing potentials (TIPs) of the different trace gases are represented in Tables 11.2, 11.3 and 11.4.

IMAGE APPROACH TIP	Time horizon is 100 years	Time horizon is 500 years
CO_2	1	1
CH_4	22	8
N_2O	320	240
CFC-11	4904	2099
CFC-12	9295	5861

Table 11.2: TIP for modelling approach with IMAGE.

ANALYTICAL APPROACH * TIP	Time horizon is 100 year	Time horizon is 500 years
CO_2	1	1
CH_4	28	6
N_2O	264	159
CFC-11	5067	1374
CFC-12	7630	2720

* indirect analytical approach (analytical equivalent to the modelling approach (from emissions to concentrations to temperature effect)

Table 11.3: TIP for indirect analytical approach.

ANALYTICAL APPROACH ** TIP	Time horizon is 100 years	Time horizon is 500 years
CO_2	1	1
CH_4	19	4
N_2O	336	143
CFC-11	5853	1588
CFC-12	8814	3143

** direct analytical approach (from emissions to temperature effect)

Table 11.4: TIP for analytical and modelling approaches.

From Tables 11.2, 11.3 and 11.4 it follows that, considering a CO_2 time horizon of 100 years, the results of both analytical methods correspond reasonably with the IMAGE results;.

Comparing the analytical and modelling TIP values for a CO_2 time horizon of 500 years reveals a structural difference between these two procedures. The modelling TIP values appear to be considerably higher than the analytical TIPs. This is due to the linear increase in CO_2 contribution in the analytical method (illustrated in Tables 11.3 and 11.4), in contrast to the more realistic, nonlinear way CO_2 is modelled in IMAGE, shown in Table 11.2.

Consequently, the time horizon (in fact the chosen atmospheric lifetime of CO_2) appears to be of crucial importance. However, CO_2 does not have a specific atmospheric residence time. This dynamical feature of the TIP is clearly demonstrated in Figure 11.3, giving the TIP of CH_4 as a function of the time horizon, which has been varied from 1 to 500 years. In particular, when varying the time horizon of CO_2 from 1 to 100 years, the TIP value of CH_4 sharply decreases.

11.4. RESULTS

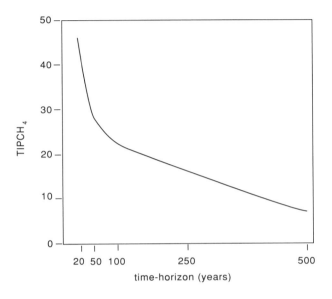

Figure 11.3: TIP of CH_4 as a function of the chosen time horizon.

Another advantage of the modelling approach is the possibility of taking the CH_4-CO-OH interactions into account. From series of experiments with CO emission impulses, it follows that a CO emission increase of 50% can increase the CH_4 concentration by about 25%.

Based on the foregoing results, an uncertainty range of TIPs for the most prominent greenhouse gases is presented in Table 11.5 for a time horizon of 100 years. This range is compared with the figures of IPCC (1990).

TIP	TIP range based on this study for a time horizon of 100 years	TIP values of IPCC (1990) for a time horizon of 100 years
CO_2	1	1
CH_4	19–28	21
N_2O	264–336	290
CFC-11	4904–5853	3500
CFC-12	7630–9295	7300

Table 11.5: TIP approach of IMAGE (with time horizon of 100 years) compared to the TIP approach followed by Lashof and Ahuja (1990)

Generally, the TIPs presented here, are somewhat higher than the (tentative) TIPs or GWPs of the IPCC, especially for the CFCs.

Using the Temperature Increasing Potentials calculated with IMAGE the relative contributions of the different greenhouse gases for the year 1985 can be calculated. In Figure 11.4 the original relative contributions are given, based on the simulated atmospheric concentrations of IMAGE.

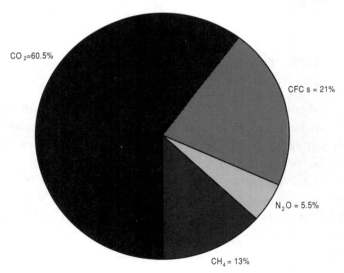

Figure 11.4: Relative contributions of trace gases to equilibrium temperature increase for the year 1985, based on global atmospheric concentrations calculated with IMAGE.

11.4. RESULTS

Figures 11.5 and 11.6 give relative cumulative contributions of the major greenhouse gases, which are based on emissions in 1985, combined with the TIP ratios calculated with IMAGE, assuming a time horizon of 100 years.

Figure 11.5 shows the contributions of the different greenhouse gas emissions of 1985 to the cumulative equilibrium temperature effect over the next 100 years. The resulting relative contributions of Figure 11.5 consist of the 1985 emissions multiplied by the TIP ratios of IMAGE. In Figure 11.6 the same procedure is followed, but only for the man-made (or anthropogenic) emissions. Comparing these figures, the TIP calculations of Figure 11.5 indicate an underestimation of CH_4 and N_2O as greenhouse potentials, and on the other hand an overestimation of CFC-11 and CFC-12. The minor role of CFCs according to the IMAGE TIP concept can be clarified by the realization of the Montreal Protocol, by which the emissions of these gases will be sharply reduced (United Nations Environment program, 1987). The higher contribution of Figure 11.4 is based on the current rapidly increasing concentrations of these CFCs. Recently other CFCs and also halons have been incorporated in IMAGE, allowing the TIPs of these greenhouse gases to be estimated with IMAGE, see Lashof and Rotmans (1990).

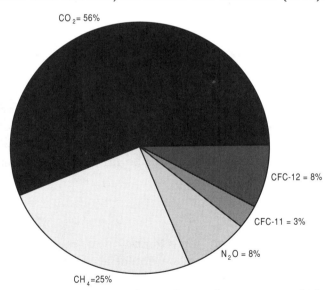

Figure 11.5: Relative contributions of greenhouse gas emissions of the year 1985 to cumulative equilibrium temperature increase for the next century, following the TIP concept according to IMAGE simulations.

Taking into account only anthropogenic emissions, the contribution of CO_2 becomes about 66%, as is shown by Figure 11.6. It should be noted that the CO_2 emission is composed of the fossil fuel component, about 5.4 GtC, and the deforestation component, about 1.5 GtC according to Swart and Rotmans (1989a,b).

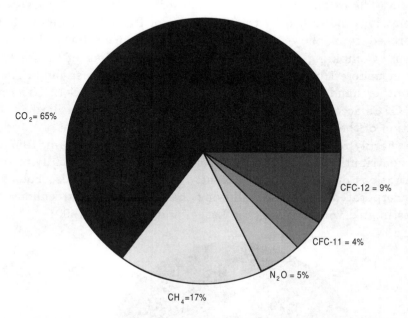

Figure 11.6: Relative contributions of man-made (or anthropogenic) greenhouse emissions of the year 1985 to cumulative equilibrium temperature increase for the next century, following the TIP concept according to IMAGE simulations.

11.5 Conclusions

Although surrounded by many uncertainties it is possible to estimate an index by which the global temperature potential of various greenhous gases can be compared. Such an index is the Temperature Increasing Potential, or TIP, which has been developed in different ways. In determining the TIP, both analytical and simulation methods can be applied, and although there appeared to be differences, both methods can be used in calculating TIP values.

Simulation experiments with IMAGE demonstrate the dynamical aspect

11.5. CONCLUSIONS

of the TIP. Therefore the time-dependent TIP definition, presented here as calculating at any time the temperature effect of a 1 (or 10^{-3}) Gt emission of a particular greenhouse gas compared to that of CO_2, is to be preferred to the static TIP definitions given so far.

Results based on calculations with IMAGE showed that particularly CH_4 and N_2O, being about 22 and 320 times as effective as CO_2 respectively, might now be underestimated. This appeared from the relative contributions of greenhouse gases of 1985 to global warming over the next century, calculated with the TIP concept according to IMAGE, which showed a considerable share of CH_4 and N_2O especially. Based on the TIP estimates it can be concluded that, next to CO_2, CH_4 and N_2O will be threatening greenhouse potentials for the future. To a lesser degree the same holds for CFC-11 and CFC-12, although the emissions of these both gases will be sharply reduced by the realization of the Montreal Protocol.

With these TIPs, for each trace gas future global temperature increases, based on emission potentials, can be estimated directly. In this way these TIPs can be used to define quantified environmental targets which can serve as reference values for the development of international response strategies. While using TIPs as a basis for comprehensive regulation of greenhouse gas emissions may not be workable in practice, TIPs can serve as an important instrument in relating emissions to environmental targets, at least on the conceptual level.

11.6 Appendix

In the analytical approach for each trace gas two equilibrium temperature increase effects are calculated, both with and without an impulse of 1 Gt.

CO_2

Following Wigley (1987) the equilibrium temperature effect due to CO_2 can be defined as:

$$T_{CO_2}(t) = T_{2xCO_2}/Ln(2) * Ln(pCO_2(t)/pCO_2(0)) \qquad (11.10)$$

with:
$T_{CO_2}(t)$ = equilibrium temperature increase due to CO_2 at time t (in $°C$)
T_{2xCO_2} = temperature increase for a doubled CO_2 concentration (in $°C$)
$pCO_2(0)$ = initial CO_2 concentration, in the year 1985 (in ppm)

To calculate the temperature effect over a longer period, Equation (11.10) has to be integrated:

$$T_{CO_2} = \int_0^T TCO_2^i(\tau)d\tau - \int_0^T TCO_2^{wi}(\tau)d\tau \qquad (11.11)$$

where:
TCO_2^i and TCO_2^{wi} are the temperature effect of CO_2 with and without impulse and T is the time horizon.

However, assuming an immediate steady state concentration of CO_2, the temperature effect of CO_2 without impulse can be neglected. Thus, combining (11.4), (11.10) and (11.11) yields expression (11.5).

CFCs

The CFC temperature effect is defined according to Ramanathan et al. (1985):

$$TCFC(t) = ACFC * [pCFC(t) - pCFC(0)] \qquad (11.12)$$

with:
$TCFC(t)$ = equilibrium temperature increase due to CFC (in $°C$)
$ACFC$ = CFC temperature coefficient, obtained from Ramanathan et al. (1985) (in $°C/ppb$)

11.6. APPENDIX

Then the total temperature effect, integrated over the time horizon T is:

$$TCFC = \int_0^T TCFC^i(t)dt - \int_0^T TCFC^{wi}(t)dt \qquad (11.13)$$

where $TCFC^i(t)$ and $TCFC^{wi}(t)$ are the temperature effect at time t with and without an emission impulse.

Combining (11.6), (11.12) and (11.13) yields for the temperature effect with impulse:

$$\begin{aligned} TCFC^i \ = \ & ACFC * [(1 - e^{-T/LFTCFC}) * \\ & (pCFC(0) + CVCFC * EMCFC) \\ & * LFTCFC - LFTCFC * PCFC(0) \end{aligned} \qquad (11.14)$$

and without emission impulse:

$$\begin{aligned} TCFC^{wi} \ = \ & ACFC * [(1 - e^{-T/LFTCFC}) * pCFC(0) * LFTCFC \\ & - LFTCFC * PCFC(0)] \end{aligned} \qquad (11.15)$$

where $TCFC^i$ and $TCFC^{wi}$ are the total temperature rises with and without emission impulse respectively, and T is the time horizon. Equations (11.14) and (11.15) produce the total temperature rise, given in (11.7).

CH$_4$ and N$_2$O

The equilibrium temperature effect of CH$_4$ is based on model results of Kiehl and Dickinson (1987):

$$TCH_4 = ACH_4 * (\sqrt{pCH_4(t)} - \sqrt{pCH_4(0)}) \qquad (11.16)$$

with:
$TCH_4(t)$ = equilibrium temperature increase due to CH$_4$ (in $^\circ C$)
$pCH_4(t)$ = atmospheric CH$_4$ concentration (in ppm)
$pCH_4(0)$ = initial atmospheric CH$_4$ concentration, in 1985 (in ppm)
ACH_4 = CH$_4$ temperature coefficient, obtained from Kiehl and Dickinson (1987)

Then the total temperature effect with an emission impulse can be described as:

$$TCH_4^i = ACH_4 * e^{(\delta/LFTCH_4+n)} * [\sqrt{(CVFCH_4 * EMCH_4 + PCH_4(0)) * 2 * (LFTCH_4+n)}$$
$$(1 - e^{-T/2*(LFTCH_4+n)}) - \sqrt{(pCH_4(0) * (LFTCH_4+n)}]$$
(11.17)

with:
$CVFCH_4$ = conversion factor of CH_4 (in ppm/Gt)
$LFTCH_4$ = atmospheric lifetime of CH_4 (in years)
$EMCH_4$ = emission impulse of 1 Gt CH_4 (in Gt)
T = time horizon

and without this emission impulse:

$$TCH_4^{wi} = ACH_4 * e^{(\delta/LFTCH_4+r)} * [\sqrt{PCH_4(0)} * 2$$
$$* (LFTCH_4+r) * (1 - e^{-T/2*(LFTCH_4+n)})$$
$$- \sqrt{PCH_4(0)} * (LFTCH_4+r)]$$
(11.18)

Then the total temperature effect, integrated over the atmospheric lifetime of CH_4 is obtained by subtracting the temperature effect without emission impulse from the temperature with emission impulse, yielding Equation (11.9).

The corresponding formula for N_2O, given in Equation (11.8), can be derived identically.

Chapter 12

Sensitivity Analysis

12.1 Introduction

The technique of metamodelling and experimental design (Kleijnen, 1987), which has already been applied to IMAGE (Rotmans et al., 1988, Rotmans and Vrieze, 1990, Rotmans et al., 1990), is used to perform sensitivity experiments with IMAGE. In this chapter it is explained how the technique of metamodelling enables us to search for a relationship between input and output variables of IMAGE and how experimental designs can be used to carry out experiments on this model in an efficient and effective way. Various modules of IMAGE have been analysed in this way; the modules concerning the costs of dike raising and the carbon cycle module. First the modules concerning dike raising and the costs of dike raising have been analyzed, because these modules are expected to behave linearly. Afterwards the carbon cycle module was subject to sensitivity analysis, since this is a quintessential constituent of the model. The carbon cycle module is split up into an ocean module and a terrestrial biota module, as described in Chapter 3, and these modules are treated separately. The deforestation module, already integrated in the carbon cycle module now, has not yet been put to the sensitivity test. Moreover it is worth mentioning that an uncertainty analysis has been applied on the modules concerning dike raising and the ocean module of the carbon cycle, with help of the Latin Hypercube Sampling method (Lammerts, 1989).

12.2 Metamodelling

Following Kleijnen (1987), a metamodel has to be interpreted as a regression model of the actual simulation model. Suppose the functional relationship between an independent variable, denoted by Y, and k chosen factors z of the simulation model is given by:

$$Y = f_1(z_1, z_2, \ldots, z_k) \qquad (12.1)$$

where $z_1, \ldots, z_k$ correspond to the chosen factors of the simulation model. Using a Taylor expansion we can approximate (12.1) by

$$\tilde{Y} = \sum_{j=0}^{n} a_j w_j \simeq Y \qquad (12.2)$$

where n is some natural number, $a_j, j \in \{0, \ldots, n\}$ are constants and $w_j, j \in \{0, \ldots, n\}$ are suitable chosen functions of $z_1, \ldots, z_k$ with $w_0 = 1$. For instance a first-order approximation would give $n = k$ and $w_j = z_j, j \in \{0, \ldots, k\}$. Alternatively an approximation with interactions would give $n = k + (k*(k-1))/2$, where $w_0, \ldots, w_k$ are the same as in the first-order case and $w_{k+1}, \ldots, w_{k+(k*(k-1))/2}$ reflect first-order interactions. Notice that from a theoretical viewpoint there is no restriction on the functional relationship between the w_k's and the z_k's; they may even be logarithmically related.

Next the coefficients $a_0, \ldots, a_n$ in the metamodel are estimated by linear regression, using ordinary least squares, and it is tested whether the metamodel is indeed valid. If not, the metamodel might be improved by adding more terms to (12.2). Now let Y_i denote the resulting output in run i where the levels of the factors are $z_{1i}, \ldots, z_{ki}$ which determine $w_{1i}, \ldots, w_{ni}$. Let m be the number of runs, $m \geq q = n + 1$, where q is the number of constants a_j, and let W be the matrix

$$W = \begin{array}{cccc} 1 & w_{11} & \ldots & w_{1n} \\ 1 & w_{i1} & \ldots & w_{in} \\ 1 & w_{m1} & \ldots & w_{mn} \end{array} \qquad (12.3)$$

Then it is well known, that

$$\hat{\beta} = (W'W)^{-1} W' \bar{Y}' \qquad (12.4)$$

12.2. METAMODELLING

is the least squares estimator of the vector of coefficients $(a_0, a_1, \ldots, a_n)$, where $\bar{Y} = (Y_1, \ldots, Y_m)$. Of course in (12.4) it is assumed that $(W'W)$ is a non-singular matrix. Equation (12.4) yields the following estimator (denoted by $\hat{Y}$) of $\tilde{Y}$ in (12.2):

$$\hat{Y} = w(W'W)^{-1}W'\bar{Y}', \tag{12.5}$$

where w is the vector of functionals $(w_0, w_1, \ldots, w_n)$. Since $\tilde{Y}$ is an approximation of Y it follows that $\hat{Y}$ is also an approximation of Y. Two types of errors have been introduced:

1. The choice of $\tilde{Y}$. This choice, determining the metamodel, may lead to a systematic error. The validition procedure has to prove the correctness of this choice.

2. The choice of $z_{i1}, \ldots, z_{ik}$, with $i = 1, \ldots, m$ in the estimation procedure of $a_0, a_1, \ldots, a_n$, which can be interpreted as "noise".

This mathematical analysis can be refined if a statistical (sub)model is added for the fitting errors $e = Y - \hat{Y}$ (Rotmans et al (1990), Kleijnen et al. (1990)). If it is assumed that these errors are normally and independently distributed with common variance, say σ^2. Then the least squares algorithm yields the Best Linear Unbiased Estimator (BLUE); The variances of these estimators are on the main diagonal of the variance-covariance matrix of $\hat{B}$:

$$cov(\hat{B}) = (X'X)^{-1}\sigma^2 \tag{12.6}$$

Recently a group of Americans statisticians proposed a more general model. They assume that the errors e are not independent but form a stationary process with a specific auto-correlation function, see Sachs et al. (1989), Kleijnen (1990). σ^2 can be estimated by the mean squared residuals:

$$\hat{\sigma}^2 = \sum_{i=1}^{m}(Y_i - \hat{Y}_i)/(m - n - 1) \tag{12.7}$$

where $\hat{Y}_i$ denotes the predicted observation for run i, using (12.5). To test whether a coefficient a_k in (12.2) significantly differs from 0 the t statistic

$$t_{d.k} = \hat{\beta}_k/s_k, \quad k = 0, 1, \ldots, n \tag{12.8}$$

can be used, where s_k, the standard deviation of the estimator $\hat{B}_k$, equals the corresponding main diagonal element of the covariance matrix of $\hat{\beta}$: $(W'W)^{-1}\hat{\sigma}^2$. Further, d stands for the degrees of freedom of t, which equals the degrees of freedom of $\hat{\sigma}^2$: $d = m - n - 1$. When validating deterministic simulation models, Kleijnen (1987) discourages the application of the studentized deviation $(Y_{m+1} - \hat{Y}_{m+1})/(\widehat{var}(\hat{Y}_{m+1}) + \widehat{var}(Y_{m+1}))$ where $m+1$ refers to an extra run. Instead he recommends the use of the relative prediction error

$$RP_{m+1} = \hat{Y}_{m+1}/Y_{m+1} \qquad (12.9)$$

where RP_{m+1} is the relative prediction error of the extra run, $\hat{Y}_{m+1}$ is the predicted observation of the extra run, using formula (12.5), and Y_{m+1} is the observed outcome of the simulation program. Though there exists a method of approximating the variance of the ratio of two random variables (Efron and Gong, 1983, pag. 40), a simple statistic for this relative prediction error is not known. In addition the power of such a statistical test becomes small if the model is bad. Therefore Kleijnen (1987) suggests rejecting the regression model if these errors are "too big", for example bigger than 5% (this number depends on the distribution of the errors throughout the model, on the nature of the simulation model etc.).

12.3 Experimental Design

Experimental design is a statistical technique for doing experiments in an effective and efficient way. The input variables are changed in a systematic way to discover the behaviour of the output variable(s). With respect to metamodels experimental design involves the setting of factors in each simulation run to estimate $\hat{\beta}$. Here an experimental design is identified by a given set of values for the factors $z_1, \ldots, z_k$. In experimental designs these factors have only a limited number of levels, mostly two. An experimental design is effective if all relevant coefficients can be estimated, and efficient if variances of the estimators are minimal compared to the estimators of designs with the same number of experiments. Or, in other words, if variances are equal, the method which requires the least number of runs is the most efficient.

In order to minimize the number of runs with the simulation model in estimating the regression coefficients, several well-known experimental designs are used.

Assuming a first-order approximation, with $n = k$ and $w_j = z_j, j \in \{0, \ldots, k\}$ then (12.2) can be written as:

$$\tilde{Y} = \sum_{j=0}^{k} a_j z_j \qquad (12.10)$$

12.3. EXPERIMENTAL DESIGN

In this case an experimental design can be based on 2 levels per factor, namely the highest and lowest value of each factor. A complete factorial design, where each combination of values occurs, would take 2^k experiments or runs. However, by associating p factors with irrelevant interactions (products of factors) the number of experiments needed is reduced to 2^{k-p} (Kleijnen, 1987, Rotmans et al., 1988).

By applying a simple linear transformation the factors can be normalized such that the highest and lowest value of each factor is respectively $+1$ and -1 (Bettonvil and Kleijnen, 1988, Kleijnen, 1987, see section 12.4.3). Such a normalization can be carried out very simply by the substitution:

$$z_j = c_j x_j + d_j \qquad (12.11)$$

where $c_j = (H_j - L_j)/2$ and $d_j = (H_j + L_j)/2$; H_j and L_j refer to the highest and lowest value of factor z_j, respectively. Then Equation (12.10) transforms into:

$$\tilde{Y} = \sum_{j=0}^{k} \alpha_j x_j \qquad (12.12)$$

where $\alpha_0, \ldots, \alpha_k$ follow straightforwardly as functions of $a_0, \ldots, a_k$ by substitution of (12.11) into (12.10). Since each factor in (12.12) can vary between $+1$ and -1, the coefficients $b_j, j \in \{0, \ldots, k\}$ can be interpreted as $1/2$ times the effect of factor x_j when this factor changes from its lowest value to its highest one. Moreover b_0 can be interpreted as the overall (or mean) effect, averaged over all possible combinations of values of the factors. Another advantage of this normalization is that each factor now changes over the same interval $[-1; +1]$ allowing the estimators of the coefficients to be compared directly in determining the relative importance of the related factors.

To estimate quadratic terms one needs more than two values per factor. In that case 3 values per factor can be used, in a 3^{k-p} factorial design, and even five factors per factor, in a central composite design. Both the 3^{k-p} and central composite designs are described extensively in Box and Hunter (1978), Montgomery (1984), and Rotmans et al. (1988). Rotmans and Vrieze (1990) conclude that central composite designs are more appropriate than 3^{k-p} designs, because the latter imply a number of runs growing inadmissibly when $k - p$ increases.

12.4 A Metamodel for the Costs of Dike Raising

12.4.1 Introduction

In this section metamodels and experimental designs are developed for the modules concerning dike raising and the costs of dike raising for the Netherlands. In these modules dike raising as well as the costs of dike raising are calculated given the simulated sea level rise in Chapter 8. In total nine different metamodels have been formulated, calculated with nine experimental designs. Here only the two most relevant designs and metamodels are treated. For each metamodel the parameters, which a priori supposed to be important, are presented.

Subsequently the metamodel and experimental design are formulated. Next the experiments and validation of the metamodel will be carried out; then the results are presented. Finally the results will be evaluated and conclusions will be drawn. Furthermore, in the last metamodel the effect of standardization of the factors is examined.

12.4.2 Input and output variables

The cumulative costs of dike raising (in billions of guilders) for a chosen strategy of dike raising with constant safety level (den Elzen and Rotmans, 1988), are chosen as output variable, denoted as Y. The costs of dike raising, approximated by the metamodel, is denoted as $\tilde{Y}$. Factors of which the influence on output variables has to be estimated are denoted originally as z, and in normalized form as x. The estimates of their effects are denoted by $\hat{\alpha}$, see section 12.4.3. The costs of dike raising refer to the year 2100, the chosen end-point of simulation.

12.4.3 Specification of the First Metamodel for the Costs of Dike Raising

For the modules of dike raising and the costs of dike raising the following eleven factors are selected (also see Chapter 8, and 9):

z_1 : initial costs of dike raising;
z_2 : marginal costs of dike raising per meter;
z_3 : reduced rate of interest, meaning the difference between the real rate of interest and economic growth;
z_4 : mimimal dike raising;
z_5 : gradient of flooding frequencies of storm surges,

12.4. A METAMODEL FOR THE COSTS OF DIKE RAISING

where flooding frequency is defined as the probability of once exceeding a certain height;
z_6 : natural sea level rise;
z_7 : melting of the glaciers;
z_8 : melting of the Greenland ice cap;
z_9 : net effect at the Antarctics (negative, because of the accumulation effect);
z_{10} : depth of the mixed layer of the ocean;
z_{11} : delay of the thermal expansion, representing the inertia of the ocean.

The upper and lower bounds of these factors, mainly based on national and international scientific knowledge are represented in Table 12.4.1. The basic values in this table are the 'standard' values of the parameters, as used in the model. By means of a simple linear transformation, described in Equation (12.11), the factors z_j are transformed into normalized factors x_j with two levels, -1 and $+1$

Additionally seven two-factor interactions are expected to be important, namely those between the factors x_1 and x_2, and between x_5 and x_6, x_7, x_8, x_9, x_{10} and x_{11}. So it is assumed that the simulation model can be approximated by the following regression equation:

$$Y_m(i) = \alpha_0 + \sum_{j=1}^{11} \alpha_j x_j(i) + \alpha_{1,2} x_1(i) x_2(i) + \alpha_{5,6} x_5(i) x_6(i)$$
$$+ \alpha_{5,7} x_5(i) x_7(i) + \alpha_{5,8} x_5(i) x_8(i) + \alpha_{5,9} x_5(i) x_9(i)$$
$$+ \alpha_{5,10} x_5(i) x_{10}(i) + \alpha_{5,11} x_5(i) x_{11}(i) \quad (i = 1, \ldots, m)$$
(12.13)

with:
$Y_m(i)$: estimated costs of dike raising at time 2100 in the i-th experiment, or the i-th combination of factor levels; $(i = 1, \ldots, m)$;
$x_j(i)$: value of factor x_j in the ith experiment (or combination); $(j = 1, \ldots, 11)$;
α_j : main effect of factor x_j;
α_0 : the 'overall' average value;
$\alpha_{j,j'}$: interaction between factor x_j and $x_{j'}, j \neq j'$;
m : number of experiments (runs) (or the number of combinations).

factor	minimum value (−)	basic value	maximum value (+)	dimension
z_1	1.1E+08	2.2E+08	16.5E+08	fl/m
z_2	4.4E+08	15.4E+08	26.4E+08	fl/m
z_3	1.0	2.0	3.0	%/year
z_4	0.30	0.30	0.50	m
z_5	2.5	3.0	3.5	
z_6	0.0015	0.0015	0.002	m/year
z_7	0.0009	0.0011	0.0012	$m/^{\circ}C$ year
z_8	0.000128	0.000147	0.000166	$m/^{\circ}C$ year
z_9	−0.0003648	−0.000237	−0.0001088	$m/^{\circ}C$ year
z_{10}	70	75	100	m
z_{11}	15	25	35	year

Table 12.4.1: Factors with their upper and lower bounds.

12.4.4 Experimental Design for the First Metamodel

A full factorial experimental design, or a 2^k design, where k denotes the number of factors, would require 2^k experiments. Because that is too many, another experimental design is chosen: a 2^{k-p} design (Kleijnen, 1987). Taking into account seven particular interactions between the eleven factors (see section 12.4.3), $2^{11-6} = 32$ runs must be done in order to estimate the $(1+11+7 = 19)$ αs in the metamodel. This must be executed as follows: between the eleven factors five factors are selected which occur comparatively rarely in the important interactions (at least the interactions which are supposed to be important, based on the prior knowledge and experience of the model builder); $(x_1, x_2, x_3, x_4$ and $x_6)$ are selected and for these factors a full experimental design is worked out, requiring 2^5 runs. The remaining six factors $(x_5, x_7, x_8, x_9, x_{10}$ and $x_{11})$ are written as combinations of the five main factors (Kleijnen, 1987, pp. 295–300):

$$5 = 1\ 3 \quad 7 = 1\ 4 \quad 8 = 1\ 6 \quad 9 = 2\ 4 \quad 10 = 2\ 6 \quad 11 = 4\ 6 \qquad (12.14)$$

Where the notation $5 = 1\ 3$ means $x_5(i) = x_1(i)\ x_3(i)(i = 1, \ldots, m)$. This means that the value of factor 5 in combination i equals the value of factor 1 in combination i, multiplied by the value of factor 3, etc. Therefore the main effect of factor 5 is confounded with the interaction between the factors 1 and 3, with $\hat{\alpha}_5 = \hat{\alpha}_{13}$ and $E(\hat{\alpha}_5) = \alpha_5 + \alpha_{13}$, see Kleijnen (1987), pp. 295–300. Equation (12.14) gives the generators of this experimental design.

12.4. A METAMODEL FOR THE COSTS OF DIKE RAISING

12.4.5 Results of the First Metamodel

Table 12.4.2 contains the estimated effects $\hat{\alpha}$, estimated by using ordinary least squares. In Table 12.4.2 also the values of the Student statistic t_v with $v = m - q = 32 - 19 = 13$ degrees of freedom, where q is the number of effects to be estimated see (12.3). Since we wish a type-I error of 5%, only those factors are significant for which $|t_{13}| > 2.16$. These factors will be denoted by an asterisk.

Consequently the most important factors, next to x_0 are x_2, x_4, x_5, x_6, x_7, x_9, x_{10} and x_{11}.

factor	estimated effect	Student t_{13}
x_0	2.8904	207.00*
x_1	0.0243	1.74
x_2	0.0335	2.40*
x_3	−0.0061	−0.44
x_4	−0.7863	−56.31*
x_5	−0.0993	−7.11*
x_6	−0.0658	−4.71*
x_7	−0.0772	−5.53*
x_8	−0.0147	−1.05
x_9	−0.0996	−7.13*
x_{10}	−0.0474	−3.40*
x_{11}	0.0545	3.90*
$x_1 x_2$	0.0113	0.81
$x_5 x_6$	0.0023	0.16
$x_5 x_7$	−0.0075	−0.54
$x_5 x_8$	0.0001	0.01
$x_5 x_9$	−0.0062	−0.44
$x_5 x_{10}$	0.0002	0.01
$x_5 x_{11}$	0.0026	0.18

Table 12.4.2: Factors with estimated effects and t-values.

12.4.6 Further Analysis after the First Metamodel

From Table 12.4.2. it follows that α_6, α_7 and α_9 are significant and negative, implying a reduction in the costs of dike raising were the sea level to continue rising. However this is inconsistent with simulation results, showing increasing costs of dike raising with continuing sea level rise. Therefore the metamodel gives no adequate approximation of the simulation model and should be rejected.

After the first metamodel another seven metamodels have again been specified and calculated. For brevity's sake the results of these metamodels are omitted. Thoroughly analysing the simulation model teaches us that the simulation should be split up. Firstly the dike raising at a constant safety level must be calculated, using the dike raising module. Thereafter these dike raising figures are put into the module which calculates the costs of dike raising, whereupon subsequently the costs of dike raising are simulated. So far the costs of dike raising have been calculated without adapting dike raising.

A new factor is introduced, namely the safety factor (x_{12}), which, multiplied by the "Delta norm", represents the maximal permissible probability of inundation. Since the reduced rate of interest and the economic growth are directly related to each other (economic growth = nominal rate of interest - reduced rate of interest), taking only one of these three factors into account will suffice. Henceforth instead of the reduced rate of interest (x_3) only the economic growth will be considered. This factor has a basic value of 2% per year in the former experiments of Table 12.4.1. Varying this factor, from its lowest value (1% per year) to its highest value (3% per year), it appears to dominate the other factors, which indeed might be expected. The longer one delays dike raising, the more "expensive" a dike raising will be, caused by inflation. In order to estimate the other factors in a better way the economic growth factor (x_{13}) will be set to zero. In paragraph 12.4.7. a survey of all factors is given.

12.4.7 Specification of the Final Metamodel for the Costs of Dike Raising

Considering the conclusions of the last paragraph, a new series of experiments will be executed, once again with a 2^{k-p} design, but with restricted bounds, consequently causing a shrinking domain of the metamodel, see Table 12.4.3.

Following Kleijnen (1987, pp. 197–200) restricting the metamodel to a smaller domain, or equivalently, reducing the ranges of the factors is one of the possibilities to improve a false or non-valid metamodel. Once again the factors z_j are normalized to x_j by means of the linear transformation, described in Equation (12.11). After the previous experiments, it is now assumed that eight interactions are important, namely those between the factors x_1 and x_2, x_2 and x_7, x_2 and x_9, x_5 and x_6, x_5 and x_7, x_5 and x_8, x_5 and x_9, and x_5 and x_{10}. The following generators are chosen:

12.4. A METAMODEL FOR THE COSTS OF DIKE RAISING

$$2 = 4\ 12\quad 5 = 1\ 4\quad 7 = 1\ 6\quad 9 = 1\ 8\quad 10 = 6\ 8\quad 11 = 8\ 12 \qquad (12.15)$$

This results in Table 12.4.4 (with $m = 32$ and $q = 20$).

factor	minimum value (−)	basic value	maximum value (+)	dimension
z_1	1.65E+08	2.2E+08	2.42E+08	fl/m
z_2	9.9E+08	15.4E+08	20.9E+08	fl/m
z_4	0.30	0.30	0.40	m
z_5	2.5	3.0	3.5	
z_6	0.00125	0.0015	0.00175	m/year
z_7	0.0010	0.0011	0.0012	m/°C year
z_8	0.000128	0.000147	0.000166	m/°C year
z_9	−0.0003648	−0.000237	−0.0001088	m/°C year
z_{10}	65	75	80	m
z_{11}	20	25	30	year
z_{12}	1.375	1.5	1.5	

Table 12.4.3: Revised factors and their upper and lower bounds.

factor	estimated effect	Student t_{12}
x_0	2.1273	144.28*
x_1	0.1596	10.83*
x_2	0.5693	38.61*
x_4	−0.0815	−5.53*
x_5	−0.0093	−0.63
x_6	0.1482	10.05*
x_7	0.0730	4.95*
x_8	0.0415	2.81*
x_9	0.2009	13.63*
x_{10}	0.0718	4.87*
x_{11}	−0.0323	−2.19*
x_{12}	0.0052	0.35
$x_1\,x_2$	0.0206	1.39
$x_2\,x_7$	0.0205	1.39
$x_2\,x_9$	0.0794	5.39*
$x_5\,x_6$	0.0003	0.02
$x_5\,x_7$	−0.0306	−2.07
$x_5\,x_8$	−0.0052	−0.35
$x_5\,x_9$	−0.0028	−0.19
$x_5\,x_{10}$	0.0155	1.05

Table 12.4.4: Factors with estimated effects and t-values.

12.4.8 Validation of the Final Metamodel

The estimated effects do now have the right signs. So the metamodel explains reasonably well; the question is, however, whether the metamodel does predict acceptably. To answer this question the metamodel will be validated in four different ways:

1. The estimated costs from the metamodel ($\hat{Y}$) are compared to the costs calculated by the simulation model (Y). Next, for each run the relative residual (in terms of percentage) is calculated:

$$\begin{aligned}\text{rel. residual} &= 100 * (\text{ simulated costs} \\ &\quad - \text{ estimated costs})/\text{simulated costs} \\ &= 100 * (Y - \hat{Y})/Y \end{aligned} \quad (12.16)$$

This yields Table 12.4.5. The relative residuals in this table are less than 10%, which is acceptable, in light of the uncertainties involved in these modules.

2. Cross-validation (Kleijnen, 1987, pp. 188–190) is a technique where a number of times (six in this case) one run is left out from the experimental design. Next the whole estimation procedure is executed with the $(m-1) = (32-1)$ remaining runs and $\hat{Y}$ (outcome of a metamodel) of the omitted run is predicted. Thereafter $\hat{Y}$ is compared to Y (outcome of a simulation model) of the omitted run. In Table 12.4.6 the results of cross-validation are given. The results of this cross-validation are satisfactory.

3. A number of extra runs is done, again six. These extra experiments are based upon new combinations of values of the input variables, according to a random experimental design: the + and − values are drawn with chance 0.5. Based on the estimates of the first 32 runs the metamodel can calculate the value $\hat{Y}(m+1)$. This value is compared to the $(m+1)$th outcome of the simulation model, etc. The extra runs and the results of these experiments are represented in Table 12.4.7.

12.4. A METAMODEL FOR THE COSTS OF DIKE RAISING

run	estimated costs ($\hat{Y}$)	simulated costs (Y)	relative residuals $(100 * (Y - \hat{Y})/Y)(in\%)$
1	3.372	3.397	0.7
2	2.634	2.651	0.6
3	2.774	2.730	−1.6
4	2.385	2.367	−0.8
5	2.219	2.233	0.6
6	1.829	1.752	−4.4
7	1.644	1.548	−6.2
8	1.480	1.618	8.5
9	2.046	2.020	−1.3
10	1.497	1.482	−1.0
11	1.531	1.577	2.9
12	1.331	1.347	1.2
13	3.653	3.699	1.2
14	2.816	2.832	0.6
15	2.996	3.032	1.2
16	2.385	2.307	−3.4
17	2.261	2.317	2.4
18	2.686	2.712	1.0
19	2.101	2.126	1.2
20	2.751	2.662	−3.3
21	1.596	1.601	0.3
22	1.650	1.623	−1.7
23	1.168	1.181	1.1
24	1.571	1.600	1.8
25	1.417	1.300	−9.0
26	1.396	1.430	2.4
27	1.175	1.210	2.9
28	1.378	1.406	2.0
29	2.547	2.543	−0.2
30	2.791	2.816	0.9
31	2.202	2.187	−0.7
32	2.794	2.767	−1.0

Table 12.4.5: Estimated costs and simulated costs and relative residuals.

omitted run	Y	$\hat{Y}$	$100 * (Y - \hat{Y})/Y (in\%)$
1	3.397	3.329	1.97
2	2.651	2.606	1.69
3	2.730	2.848	−4.32
4	2.367	2.416	−2.04
5	2.233	2.196	1.67
6	1.752	1.956−	−11.63

Table 12.4.6: Cross-validation

extra run	Y	$\hat{Y}$	$100 * (Y - \hat{Y})/Y (in\%)$
33	1.5765	1.546	1.93
34	2.6363	2.483	5.81
35	1.6123	1.738	−7.79
36	1.8543	1.918	−3.44
37	1.3761	1.306	5.08
38	1.3761	1.406	−2.14

Table 12.4.7: Extra runs.

Table 12.4.7 shows that the relative residuals do not deviate much from zero, and are thus acceptable.

4. Finally two extra validation runs are executed, where points are chosen within the empirical area, instead of at the corners. In these runs the basic values from Table 12.4.3 have been standardized to values between −1 and +1 and substituted in the regression equation, yielding a $\hat{Y}$ value of 2.1969. The accompanying value Y is 1.7872. Thus the relative residual $100 * (Y - \hat{Y})/Y$ is −22.92%. The same procedure has been adopted for the second run, with values "close" to the basic values:
$x_1 = 2.0E + 08$, $x_2 = 1.0E + 09$, $x_4 = 0.350$, $x_5 = 2.75$, $x_6 = 0.0012$, $x_7 = 0.00105$, $x_8 = 0.0014$, $x_9 = -0.0003$, $x_{10} = 67.5$, $x_{11} = 22.5$, and $x_{12} = 1.4$.

This yields a $\hat{Y}$ of 1.2413 and a Y of 1.1689; the relative residual of this run is −6.19%. Because the relative residuals in the Tables 12.4.5, 12.4.6, and 12.4.7 and in the second extra run are less than 10%, except for the sixth run (see Table 12.4.6), this metamodel is accepted.

12.4.9 Scaling Effects

In order to determine the influence of factor standardization on the results, a couple of inverse transformations have been been carried out following Kleijnen (1987, pp. 341–345) and Bettonvil and Kleijnen (1988). Such an inverse transformation involves the conversion of the normalized factors x_j back to the original factors z_j. Considering only the significant factors and interactions (see Table 12.4.4), then the equation of the final metamodel, but now expressed in the original factors z_j, becomes (see section 12.4.7):

$$\begin{aligned} Y = {} & \beta_0 + \beta_1 z_1 + \beta_2 z_2 + \beta_4 z_4 + \beta_6 z_6 + \beta_7 z_7 + \beta_8 z_8 + \beta_9 z_9 + \beta_{10} z_{10} \\ & + \beta_{11} z_{11} + \beta_{2,9} z_2 z_9 \end{aligned} \quad (12.17)$$

An alternative is the "centered" model, where $\bar{z}_j = b_j = (H_j + L_j)/2$.

$$\begin{aligned} Y = {} & \delta_0 + \delta_1(z_1 - \bar{z}_1) + \delta_2(z_2 - \bar{z}_2) + \delta_4(z_4 - \bar{z}_4) + \delta_6(z_6 - \bar{z}_6) + \\ & \delta_7(z_7 - \bar{z}_7) + \delta_8(z_8 - \bar{z}_8) + \delta_9(z_9 - \bar{z}_9) + \delta_{10}(z_{10} - \bar{z}_{10}) + \\ & \delta_{11}(z_{11} - \bar{z}_{11}) + \delta_{2,9}(z_2 - \bar{z}_2)(z_9 - \bar{z}_9) \end{aligned} \quad (12.18)$$

Bettonvil and Kleijnen (1988) treat the general relationship between α, β and δ. Table 12.4.8 gives the results for the different scales. The effects order of the factors, decreasing from high to low effects, is for α:

x_2, x_9, x_1, x_6, x_4, x_7, x_{10}, x_8, x_{11}; and for β and δ:
x_9, x_8, x_7, x_6, x_4, x_{11}, x_{10}, x_1, x_2.

The interaction 2,9 appears to be relatively unimportant. The order for the parameters β and δ is exactly the same; (from Table 12.4.8 it follows that, apart from β_0 and δ_0, β and δ are identical) however the inverse transformation order (β and δ) deviates from that of the standardized model (α). To determine the correct order, the one-factor-at-a-time procedure (Kleijnen, 1987) has been applied for the factors x_2 and x_8. Setting all factors to their basic value, except for factor x_2, which is set to its maximal value, results in increased costs of dike raising from Dfl. 1.7872 billion to Dfl. 2.2684 billion; following the same procedure for x_8 instead of x_2 yields an increase in costs of dike raising from Dfl. 1.7872 billion to Dfl. 1.8011 billion. Apparently x_2 dominates x_8, so the standardized model gives the correct priority order. This is in line with the general conclusion in Kleijnen (1987) and Bettonvil and Kleijnen (1988).

index j	standardized effect (α_j)	original effect (β_j)	centered effect (δ_j)
0	2.1273	−1.9140	2.1273
1	0.1596	0.0000	0.0000
2	0.5693	0.0000	0.0000
4	−0.0815	−1.6300	−1.6300
6	0.1482	592.7999	592.7999
7	0.0730	730.0000	730.0000
8	0.0415	2184.2104	2184.2104
9	0.2009	1569.5016	1569.5016
10	0.0718	0.0096	0.0096
11	−0.0323	−0.0065	−0.0065
2,9	0.0794	0.0000	0.0000

Table 12.4.8: Three alternative models.

12.4.10 Conclusions

The first metamodel, estimated by a 2^{k-p} experimental design, has been rejected, because it did not adequately reflect the simulation model of dike raising costs: the signs of a number of estimated parameters were wrong. Next, after a series of nine experiments with various kinds of metamodels, a final metamodel has been specified and worked out, again estimated with a 2^{k-p} experimental design, with a smaller domain (the factor bounds have been narrowed). After having validated the metamodel extensively, it has been accepted. The priority order of factors influencing the costs of dike raising is, in decreasing order:

x_2 : marginal costs of dike raising per meter;
x_9 : net effect at the Antarctics (negative, because of the accumulation effect);
x_1 : initial costs of dike raising;
x_6 : natural sea level rise;
x_4 : mimimal dike raising;
x_7 : melting of the glaciers;
x_{10}: depth of the mixed layer of the ocean;
x_8 : melting of the Greenland ice cap;
x_{11}: delay of the thermal expansion, representing the inertia of the ocean; and the interaction between the factors x_2 and x_9.

12.5. A METAMODEL FOR THE OCEAN MODULE

The marginal costs of of dike raising per meter appears to be of major importance, which is according to the expectations. The order of the remaining eight significant factors as well as the interaction, contains several surprises. This leads to an increased insight in the simulation model. The parameters of the standardized metamodel give the correct order of ranking of the parameters, in contrast to the non-standardized and centered metamodels, respectively. This means that it is not necessary to carry out the inverse transformation from the standardized x_j to the originals z_j. The only thing that matters is the relative interval width of the corresponding factors. From now on, in the further analysis of simulation models only standardized metamodels will be considered.

12.5 A Metamodel for the Ocean Module

12.5.1 Introduction

The next module which is subjected to a sensitivity analysis is the carbon cycle module, including the relation between the concentration of carbon dioxide in the atmosphere, terrestrial biosphere, and ocean; see Figure 2.2. The carbon cycle which is analysed is the original version of Goudriaan and Ketner (1984); in the newest version the deforestation module as well as a different stratification of the ocean is included, as described in Chapter 3.

The techniques of metamodelling and experimental design are applied to the ocean and terrestrial biosphere modules. The ocean module is treated first. The chosen output variable is the atmospheric CO_2 concentration (in *ppm*) in the year 2100, the end point of simulation. The organization of this section is analagous to that of Section 12.4.

The ocean is divided into the following layers: at lower lattitudes a mixed layer of about 75 meters, and under this mixed layer one of about 325 meters. At higher lattitudes there is a mixed layer of about 400 meters. Under both these mixed layers the deep sea is divided into nine layers of about 378 meters each (in total 3400 m). As mentioned in Chapter 3 this stratification of the ocean is somewhat different from the new ocean module described in Chapter 3. Nevertheless the primary processes, determining the carbon concentration in the ocean, are identical:

1. carbon transport by mass flow of water, through deep ocean layers to the equator and back to the polar areas;
2. turbulent mixing between the different layers;
3. precipitation of undissolved CO_2 at the bottom of the ocean.

12.5.2 First Metamodel for the Ocean Module

The following ten factors have been selected:

z_1 : precipitation of undissolved CO_2 at the bottom of the ocean;
z_2 : diffusion coefficient;
z_3 : thickness of the warmer ocean mixed layer;
z_4 : thickness of the colder ocean mixed layer;
z_5 : total thickness of the nine deeper ocean layers;
z_6 : residence time of CO_2 in the thin (warmer) ocean surface mixed layer;
z_7 : residence time of CO_2 in the thick (colder) ocean surface mixed layer;
z_8 : massflow of water;
z_9 : ocean area; and
z_{10}: atmospheric coefficient, converting CO_2 emissions into atmospheric CO_2 concentrations

Table 12.5.1 contains the factors with their basic, upper and lower bounds.

factor	minimum value (−)	basic value	maximum value (+)	dimension
z_1	7	8	9	GtC/year
z_2	3716	4000	5984	cm^2/sec
z_3	70	75	85	m
z_4	300	400	500	m
z_5	3000	3400	4000	m
z_6	0.05	1	2	year
z_7	15	20	25	year
z_8	2.1	2.3	2.5	$10^{15} m^3$/year
z_9	0.32	0.36	0.40	$10^{12} m^2$
z_{10}	0.469	0.471	0.472	ppm/GtC

Table 12.5.1: Factors with their upper and lower bounds.

Following the same procedure as treated in section 12.4.3, the factors z_j are transformed to normalized factors x_j. Eleven interactions are expected to be important, namely between the factors x_1 and x_3, x_1 and x_5, x_2 and x_3, x_2 and x_5, x_3 and x_6, x_3 and x_7, x_5 and x_6, x_5 and x_7, x_5 and x_8, x_6

12.5. A METAMODEL FOR THE OCEAN MODULE

and x_8 and between x_7 and x_8. The number of effects to be estimated (q) is thus $1 + 10 + 11 = 22$. For a 2^{k-p} experimental design it holds that $m = 2^{k-p} > q$; thus $m = 2^{10-5} = 32$. From among the ten factors again five factors are selected which occur comparatively rarely in the important interactions (at least the interactions which are supposed to be important, based on knowledge and experience of the model-builder); x_1, x_2, x_4, x_9 and x_{10}. This leads to the following generators:

$$3 = 9\,10 \quad 5 = 4\,10 \quad 6 = 1\,9 \quad 7 = 2\,9 \quad 8 = 1\,2\,4 \qquad (12.19)$$

Executing these 32 runs and analysing the results yields Table 12.5.2.

factor	estimated effect	Student t_{10}
x_0	1183.0	67.90*
x_1	−32.2	−1.85
x_2	28.8	1.65
x_3	112.5	6.45*
x_4	316.1	18.14*
x_5	−453.0	−26.00*
x_6	−8.1	−0.47
x_7	40.5	2.33*
x_8	18.0	1.03
x_9	−12.8	−0.73
x_{10}	−72.2	−4.15*
$x_1 x_3$	−8.0	−0.46
$x_1 x_5$	5.3	0.31
$x_2 x_3$	29.0	1.66
$x_2 x_5$	−25.9	−1.49
$x_3 x_6$	−11.4	−0.66
$x_3 x_7$	10.5	0.60
$x_5 x_6$	19.2	1.10
$x_5 x_7$	−6.8	−0.39
$x_5 x_8$	−14.3	−0.82
$x_6 x_8$	−53.9	−3.09*
$x_7 x_8$	20.2	1.16

Table 12.5.2: Factors with estimated effects and t-values.

The estimated CO_2 concentration ($\hat{Y}$) can now be determined and compared to the simulated CO_2 concentration (Y). The relative residuals $100 * (Y - \hat{Y})/Y$ appear to be more than 10% in eight runs (see Table 12.5.3), which is not acceptable; so this metamodel is rejected.

run	estimated CO_2 concentration ($\hat{Y}$)	simulated CO_2 concentration (Y)	relative residuals $(100*(Y-\hat{Y})/Y)(in\%)$
1	1084	1069	-1.40
2	1032	991	-4.14
3	1107	1079	-2.59
4	1070	1067	-0.28
5	1406	1435	2.02
6	1080	1108	2.53
7	1680	1695	0.88
8	1224	1241	1.37
9	1915	1845	-3.79
10	1833	1746	-4.98
11	1900	1826	-4.05
12	1948	1890	-3.07
13	392	478	17.99
14	297	368	19.29
15	432	491	12.02
16	323	396	18.43
17	821	838	2.03
18	804	819	1.83
19	936	964	2.90
20	935	964	3.01
21	1057	1065	0.75
22	1252	1212	-3.30
23	1109	1057	-4.92
24	1174	1170	-0.34
25	2235	2317	3.54
26	2010	2061	2.47
27	2224	2287	2.75
28	2129	2224	4.27
29	527	467—	-12.85
30	667	595	-12.10
31	563	478	-17.78
32	688	616	-11.69

Table 12.5.3: Estimated and simulated CO_2 concentrations and the relative residuals.

12.5.3 Further Analysis after the First Metamodel

Thereafter five other metamodels have been specified, where the bounds of the factors x_3, x_4, x_5 and x_{10} have been varied (the domain has been broadened and narrowed, respectively). The relative residuals, however, appear to be only larger. Scaling up the lower bound of x_3 and x_{10} is the only remedy to get better results.

Therefore it was decided to carry out a new series of experiments with the help of a central composite experimental design (Rotmans et al., 1988, Kleijnen, 1987). Such a central composite design makes it possible to estimate higher order effects; for instance quadratic effects or some higher order interactions between the factors, dependent on the experimental design (Kleijnen, 1987, pag. 200).

12.5.4 Final Metamodel for the Ocean Module

To determine whether there are quadratic effects, and whether the interactions between the factors x_3 and x_4, x_3 and x_5, x_4 and x_5 and x_4 and x_7 are significant, a new metamodel is specified with the factors x_3, x_4, x_5, x_7 and x_{10}, which were significant (see Table 12.5.2). The starting-point is a 2^{5-1} experimental design with levels -1 and $+1$, with generator:

$$4 = 3\ 7\ 10 \qquad (12.20)$$

To complete the central composite design supplementary runs are added to this 2^{5-1} design, where the factors in these runs take the values 0, -2 and $+2$. The values of -2 and $+2$ are chosen arbitrarily (Rotmans et al., 1988). This makes a total of 27 runs $(16 + 2 * 5 + 1)$, leading to Table 12.5.4, where the degrees of freedom are $27 - (1 + 5 + 5 + 4) = 12$.

factor	estimated effect	Student t_{12}
x_0	1074.66	154.08*
x_3	51.77	32.07*
x_4	158.37	98.10*
x_5	−224.95	−139.35*
x_7	18.78	11.63*
x_{10}	−0.24	−0.15
x_3^2	−0.46	−0.23
x_4^2	−3.82	−1.93
x_5^2	27.78	14.05*
x_7^2	−1.24	−0.63
x_{10}^2	−0.21	−0.10
$x_3 x_4$	3.06	1.55
$x_3 x_5$	−8.35	−4.22*
$x_4 x_5$	−17.65	−8.93*
$x_4 x_7$	−12.26	−6.20*

Table 12.5.4: Factors with estimated effects and t-values.

In analogy to the validation procedure of section 12.4.8, this metamodel is validated in four different ways. The relative residuals of the runs which have already been executed are given in Table 12.5.5, while Table 12.5.6 contains the results of the cross-validation procedure. The random design and the results of the runs with this random design are given in Tables 12.5.7 and 12.5.8, respectively. All relative residuals are less than 10%, which is acceptable. The extra validation run with basic values yields a $\hat{Y}$ of 1043.58; the corresponding Y is 1055.45, which results in a relative residual of 1.12%.

12.5. A METAMODEL FOR THE OCEAN MODULE

run	estimated CO_2 concentration ($\hat{Y}$)	simulated CO_2 concentration (Y)	relative residuals $100 * ((Y - \hat{Y})/Y)(in\%)$
1	1065	1074	0.85
2	1567	1565	−0.15
3	722	719	−0.40
4	1119	1114	−0.50
5	740	741	0.16
6	1171	1162	−0.83
7	959	963	0.40
8	1428	1432	0.31
9	803	803	0.06
10	1234	1238	0.29
11	973	974	0.16
12	1441	1435	−0.46
13	1053	1053	−0.01
14	1555	1550	−0.31
15	660	669	1.42
16	1058	1054	−0.37
17	1073	1074	0.07
18	1074	1074	−0.02
19	1107	1108	−0.01
20	1032	1032	−0.05
21	1176	1177	0.08
22	969	969	−0.05
23	736	724	−1.62
24	1636	1648	0.74
25	1376	1373	−0.20
26	743	746	0.46
27	1075	1074	−0.06

Table 12.5.5: Estimated and simulated CO_2 concentrations and the relative residuals.

omitted run	Y	$\hat{Y}$	$100*(Y-\hat{Y})/Y$
3	718.58	724.49	-0.80
4	1113.79	1125.15	-1.00
6	1161.79	1183.17	-1.84
11	974.41	971.06	0.34
15	669.42	650.12	2.88
16	1053.87	1061.79	-0.75

Table 12.5.6: Cross-validation.

run	x_{10}	x_7	x_3	x_5	x_4	x_{10}	x_7	x_3	x_5	x_4	x_3x_4	x_3x_5	x_4x_5	x_4x_7
28	+	−	+	+	+	+	+	+	+	+	+	+	+	−
29	+	+	−	+	+	+	+	+	+	+	−	−	+	+
30	+	+	+	+	−	+	+	+	+	+	−	+	−	−
31	+	−	−	−	−	+	+	+	+	+	+	+	+	+
32	+	+	+	−	−	+	+	+	+	+	−	−	+	−
33	−	−	+	−	−	+	+	+	+	+	−	−	+	+

Table 12.5.7: Random experimental design.

extra run	Y	$\hat{Y}$	$100*(Y-\hat{Y})/Y$
28	1052.61	1052.2	0.04
29	974.41	972.3	0.22
30	803.20	802.2	0.12
31	1053.87	1057.3	-0.33
32	1237.55	1233.5	0.33
33	1161.79	1171.9	-0.87

Table 12.5.8: Results with the random experimental design of Table 12.5.7.

Finally the regression is carried out once more, omitting the non-significant factors x_{10}, x_3^2, x_4^2, x_7^2, x_{10}^2 and x_3x_4. The results are presented in Table 12.5.9. It can be concluded that all significant factors of Table 12.5.4. remain significant; no radical changes occur in the point estimators and therefore the metamodel is accepted.

factor	estimated effect	Student t_{18}
x_0	1068.55	503.31*
x_3	51.77	30.85*
x_4	158.37	94.36*
x_5	-224.95	-134.02^*
x_5^2	28.93	18.17*
x_7	18.78	11.19*
$x_3 x_5$	-8.35	-4.06^*
$x_4 x_5$	-17.65	-8.59^*
$x_4 x_7$	-12.26	-5.97^*

Table 12.5.9: Factors with estimated effects and t-values.

12.5.5 Conclusions

The relative residuals of the definite metamodel for the ocean module are all less than about 3%. Therefore this metamodel is accepted. The priority order of the factors in the ocean module determining the atmospheric CO_2 concentration is, in decreasing order:

x_5 : total thickness of the nine deep ocean layers;
x_4 : thickness of the colder ocean mixed layer;
x_3 : thickness of the warmer ocean mixed layer;
x_5^2 : the squared effect of the total thickness of the nine deep ocean layers
x_7 : residence time of CO_2 in the colder ocean mixed surface layer; and the interactions between the factors x_4 and x_5, x_4 and x_7, and x_3 and x_5.

12.6 A Terrestrial Biosphere Metamodel

12.6.1 Introduction

This section deals with the application of the techniques of metamodelling and experimental design on the terrestrial biosphere module of IMAGE. In the previous section an adequate metamodel for the ocean module has been developed. The latter module, together with the terrestrial biosphere module, simulates the carbon cycle module according to Goudriaan and Ketner (1984). Through metamodels an attempt has been made to elucidate the relationship between the carbon dioxide concentration in the atmosphere and in the terrestrial biosphere. Following the procedure used for the metamodel of the ocean module, the atmospheric CO_2 concentration in the year 2100 (in *ppm*) is chosen as response variable. The notation is exactly the same as in previous sections.

The terrestrial biosphere is a major reservoir in the carbon cycle. Tremendous amounts of carbon dioxide are exchanged yearly between the atmosphere and the terrestrial biosphere. As stated in Chapter 3, the crucial question with regard to the carbon cycle is whether the terrestrial biosphere is a "source" (net uptake of CO_2) or a "sink" (net release of CO_2): this question cannot yet be answered adequately. The simulated version of the terrestrial biosphere acts as a sink, based on Goudriaan and Ketner (1984). The biosphere is horizontally divided into six ecosystems: tropical forest, temperate forest, grassland, agricultural land, human area, and semi-desert area. Vertically leaves, branches, stems, roots, humus and charcoal are distinguished. Within each ecosystem carbon is transported naturally, via leaves to charcoal, in exchange with the atmosphere. Besides this natural flux mutual shifts between various ecosystems induce fluxes of carbon. Additionally human activities like burning and shifting cultivation influence largely the amount of carbon in an ecosystem (Swart and Rotmans, 1989b, Goudriaan and Ketner, 1984).

12.6.2 Various Metamodels for the Terrestrial Biosphere Module

The following 62 factors of the carbon cycle module have been selected:

z_1 : fraction charcoal from decomposition of humus;
z_2 : relative growth of the world population;
z_3 : biotic growth factor (fertilization effect, implying a biomass increase caused by the increase in the atmospheric CO_2 concentration;
z_4, z_5 and z_6 : fraction charcoal by burning of leaves, in tropical forest, temperate forest, and grassland respectively.;
$z_7, \ldots, z_{12}$: residence time of charcoal in ecosystem $1, \ldots, 6$, resp.;
$z_{13}, \ldots, z_{18}$: residence time of humus in ecosystem $1, \ldots, 6$, resp.;
$z_{19}, \ldots, z_{24}$: residence time of litter in ecosystem $1, \ldots, 6$, resp.;
$z_{25}, \ldots, z_{30}$: humification factor of ecosystem $1, \ldots, 6$, resp.;
$z_{31}, \ldots, z_{36}$: shifting cultivation in ecosystem $1, \ldots, 6$, resp.;
z_{37} : shift from tropical forest to grassland;
z_{38} : shift from tropical forest to agricultural land;

12.6. A TERRESTRIAL BIOSPHERE METAMODEL

$z_{39}, \ldots, z_{44}$: life span of leaves in ecosystem $1, \ldots, 6$, resp.;
$z_{45}, \ldots, z_{50}$: life span of branches in ecosystem $1, \ldots, 6$, resp;
$z_{51}, \ldots, z_{56}$: life span of stems in ecosystem $1, \ldots, 6$, resp; and
$z_{57}, \ldots, z_{62}$: life span of roots in ecosystem $1, \ldots, 6$, resp,

where ecosystem 1 is tropical forest, 2 is temperate forest, 3 is grassland, 4 is agricultural land, 5 is human area, and 6 is semi-desert. Initially no interactions are considered in the metamodel, because of its complexity and the many calculations necessary. The 62 main effects of this metamodel are estimated in $(2^{62-55} =)$ 128 runs, by using a so-called R IV design. Such a design yields estimates of main effects without being biased by possible interactions. In this case at least 124 runs are necessary, see Kleijnen et al. (1990). However, it appeared that $\hat{\alpha}_{62}$ was significantly positive, while this parameter was expected to be negative. Hence this metamodel is rejected.

Next the domain of the input variables is narrowed. The basic values and the upper and lower bounds of these factors are given in Table 12.6.1. Besides the 62 main effects 26 important two-factor interactions are now taken into account, namely between the factors x_1 and x_{37}, x_2 and x_{37}, x_3 and x_{37}, x_1 and x_{38}, x_2 and x_{38}, x_3 and x_{38}, x_2 and $x_{31}, \ldots, x_{36}$, x_3 and $x_{31}, \ldots, x_{36}$, x_2 and x_{44}, x_3 and x_{44}, x_5 and x_{44}, x_2 and x_{60}, x_3 and x_{60}, x_{37} and x_{60}, x_{38} and x_{60} and x_{44} and x_{60}. The "defining relation" (Kleijnen, 1987, pp. 295–301) is:

$$I =$$

$$\begin{aligned}
&= 1\ 6\ 7 & &= 2\ 6\ 8 & &= 3\ 6\ 9 & &= 4\ 7\ 8 & &= 5\ 6\ 10 \\
&= 7\ 9\ 13 & &= 8\ 9\ 14 & &= 8\ 10\ 15 & &= 9\ 10\ 16 & &= 9\ 11\ 17 \\
&= 9\ 12\ 18 & &= 11\ 12\ 31 & &= 7\ 10\ 32 & &= 8\ 11\ 33 & &= 8\ 12\ 34 \\
&= 7\ 11\ 35 & &= 7\ 12\ 36 & &= 10\ 11\ 37 & &= 10\ 12\ 38 & & \\
&= 6\ 7\ 8\ 19 & &= 6\ 7\ 9\ 20 & &= 6\ 7\ 11\ 22 & &= 6\ 7\ 12\ 23 & &= 6\ 8\ 9\ 24 \\
&= 6\ 8\ 10\ 25 & &= 6\ 8\ 11\ 26 & &= 6\ 8\ 12\ 27 & &= 6\ 9\ 10\ 28 & &= 6\ 9\ 11\ 29 \\
&= 6\ 9\ 12\ 30 & &= 6\ 10\ 11\ 39 & &= 6\ 10\ 12\ 40 & &= 6\ 11\ 12\ 41 & &= 7\ 8\ 9\ 42 \\
&= 7\ 9\ 10\ 43 & &= 7\ 9\ 11\ 45 & &= 7\ 9\ 12\ 46 & &= 7\ 10\ 11\ 47 & &= 7\ 10\ 12\ 48 \\
&= 7\ 11\ 12\ 49 & &= 8\ 9\ 10\ 50 & &= 8\ 9\ 11\ 51 & &= 8\ 9\ 12\ 52 & &= 8\ 10\ 11\ 53 \\
&= 8\ 10\ 12\ 54 & &= 9\ 10\ 11\ 56 & &= 9\ 10\ 12\ 57 & &= 7\ 8\ 10\ 60 & & \\
&= 7\ 8\ 9\ 12\ 21 & &= 7\ 9\ 10\ 11\ 44 & &= 7\ 9\ 10\ 12\ 55 & &= 6\ 7\ 8\ 9\ 58 & &= 7\ 10\ 11\ 12\ 59 \\
&= 7\ 8\ 9\ 10\ 61 & &= 7\ 8\ 9\ 11\ 62 & & & & & &
\end{aligned}$$

(12.21)

factor	minimum value (−)	basic value	maximum value (+)	dimension
x_1	0.03	0.05	0.07	
x_2	0.018	0.020	0.022	
x_3	0.25	0.5	0.75	
x_4	0.03	0.05	0.07	
x_5	0.05	0.1	0.15	
x_6	0.1	0.2	0.3	
x_7	250	500	750	year
x_8	250	500	750	year
x_9	250	500	750	year
x_{10}	250	500	750	year
x_{11}	250	500	750	year
x_{12}	250	500	750	year
x_{13}	5	10	15	year
x_{14}	25	50	75	year
x_{15}	20	40	60	year
x_{16}	15	25	35	year
x_{17}	25	50	75	year
x_{18}	25	50	75	year
x_{19}	0.5	1	1.5	year
x_{20}	1	2	3	year
x_{21}	1	2	3	year
x_{22}	0.5	1	1.5	year
x_{23}	1	2	3	year
x_{24}	1	2	3	year
x_{25}	0.2	0.4	0.6	
x_{26}	0.3	0.6	0.9	
x_{27}	0.3	0.6	0.9	
x_{28}	0.1	0.2	0.3	
x_{29}	0.25	0.5	0.75	
x_{30}	0.3	0.6	0.9	
x_{31}	5	15	25	mha/year
x_{32}	1	2	3	mha/year
x_{33}	200	400	600	mha/year
x_{34}	200	400	600	mha/year
x_{35}	0	0	1	mha/year
x_{36}	0	0	1	mha/year
x_{37}	3	6	9	mha
x_{38}	3	6	9	mha
x_{39}	0.5	1	1.5	year
x_{40}	1	2	3	year
x_{41}	0.5	1	1.5	year
x_{42}	0.5	1	1.5	year
x_{43}	0.5	1	1.5	year
x_{44}	0.5	1	1.5	year
x_{45}	5	10	15	year
x_{46}	5	10	15	year
x_{47}	5	10	15	year
x_{48}	5	10	15	year
x_{49}	5	10	15	year
x_{50}	5	10	15	year
x_{51}	25	30	35	year
x_{52}	30	60	90	year
x_{53}	25	50	75	year
x_{54}	25	50	75	year
x_{55}	25	50	75	year
x_{56}	25	50	75	year
x_{57}	5	10	15	year
x_{58}	5	10	15	year
x_{59}	0.5	1	1.5	year
x_{60}	0.5	1	1.5	year
x_{61}	5	10	15	year
x_{62}	1	2	3	year

Table 12.6.1: Factors with their upper and lower bounds.

12.6. A TERRESTRIAL BIOSPHERE METAMODEL

The effects and interactions are estimated in ($2^{62-55} =$) 128 runs. The results are given in Table 12.6.2.

factor	estimated effect	Student t_{39}	factor	estimated effect	Student t_{39}
x_0	1333.08	477.21	x_{45}	3.70	1.33
x_1	-2.93	-1.05	x_{46}	0.67	0.24
x_2	0.83	0.30	x_{47}	0.33	0.12
x_3	-100.02	-35.81	x_{48}	1.65	0.59
x_4	-3.76	-1.35	x_{49}	-0.10	-0.04
x_5	-11.92	-4.27	x_{50}	-2.17	-0.78
x_6	-11.78	-4.22	x_{51}	-2.50	-0.89
x_7	4.07	1.46	x_{52}	-6.89	-2.47
x_8	0.19	0.07	x_{53}	0.13	0.05
x_9	-3.58	-1.28	x_{54}	0.21	0.07
x_{10}	-3.71	-1.33	x_{55}	4.56	1.63
x_{11}	-2.52	-0.90	x_{56}	-0.97	-0.35
x_{12}	-5.21	-1.86	x_{57}	1.23	0.44
x_{13}	1.73	0.62	x_{58}	-0.05	-0.02
x_{14}	-0.04	-0.01	x_{59}	-0.83	-0.30
x_{15}	-25.65	-9.18	x_{60}	-0.15	-0.05
x_{16}	-7.95	-2.85	x_{61}	-1.22	-0.44
x_{17}	-5.28	-1.89	x_{62}	-0.56	-0.20
x_{18}	-3.98	-1.43	$x_1 x_{37}$	0.03	0.01
x_{19}	1.50	0.54	$x_2 x_{37}$	0.10	0.04
x_{20}	2.01	0.72	$x_3 x_{37}$	-0.61	-0.22
x_{21}	2.98	1.07	$x_1 x_{38}$	1.03	0.37
x_{22}	-3.23	-1.15	$x_2 x_{38}$	2.62	0.94
x_{23}	-1.15	-0.41	$x_3 x_{38}$	2.21	0.79
x_{24}	-0.24	-0.09	$x_2 x_{31}$	2.52	0.90
x_{25}	7.27	2.60	$x_2 x_{32}$	3.31	1.19
x_{26}	-3.82	-1.37	$x_2 x_{33}$	-0.78	-0.28
x_{27}	-38.88	-13.92	$x_2 x_{34}$	-1.99	-0.71
x_{28}	-9.63	-3.45	$x_2 x_{35}$	1.08	0.39
x_{29}	-2.88	-1.03	$x_2 x_{36}$	0.68	0.24
x_{30}	-6.11	-2.19	$x_3 x_{31}$	1.97	0.71
x_{31}	3.45	1.24	$x_3 x_{32}$	0.10	0.03
x_{32}	0.53	0.19	$x_3 x_{33}$	-0.81	-0.29
x_{33}	0.02	0.01	$x_3 x_{34}$	-0.42	-0.15
x_{34}	0.92	0.33	$x_3 x_{35}$	-0.11	-0.04
x_{35}	-0.42	-0.15	$x_3 x_{36}$	-0.03	-0.01
x_{36}	-1.68	-0.60	$x_2 x_{44}$	-0.09	-0.03
x_{37}	-11.26	-4.03	$x_3 x_{44}$	2.03	0.73
x_{38}	18.53	6.63	$x_5 x_{44}$	-0.93	-0.33
x_{39}	-0.60	-0.22	$x_2 x_{60}$	-1.00	-0.36
x_{40}	-12.08	-4.33	$x_3 x_{60}$	3.66	1.31
x_{41}	3.29	1.18	$x_{37} x_{60}$	0.35	0.13
x_{42}	-1.06	-0.38	$x_{38} x_{60}$	-0.41	-0.15
x_{43}	2.04	0.73	$x_{44} x_{60}$	-0.10	-0.04
x_{44}	2.57	0.92			

Table 12.6.2: Factors with estimated effects and t-values.

All factors now do have a correct sign. The model is validated in two different ways. The relative residuals of the runs which have already been carried out are presented in Table 12.6.3. The results of 12 extra runs are given in Table 12.6.4.

run	estimated CO_2 concentration ($\hat{Y}$)	simulated CO_2 concentration (Y)	relative residuals $100*((Y-\hat{Y})/Y)$
1	1115	1104	1.05
2	1176	1182	-4.82
3	1149	1157	-6.83
4	1254	1226	2.33
5	1193	1214	-1.77
6	1325	1317	0.63
7	1178	1165	1.05
8	1370	1395	-1.80
9	1366	1348	1.40
10	1420	1450	-2.06
11	1386	1409	-1.60
12	1453	1445	0.52
13	1419	1428	-0.64
14	1481	1453	1.91
15	1382	1363	1.39
16	1472	1484	-0.79
17	1236	1251	-1.20
18	1149	1140	0.82
19	1255	1243	0.93
20	1194	1226	-2.61
21	1265	1246	1.59
22	1229	1236	-0.52
23	1235	1245	-0.81
24	1238	1215	1.89
25	1450	1462	-0.83
26	1391	1367	1.71
27	1486	1470	1.10
28	1421	1422	-0.09
29	1453	1446	0.44
30	1380	1405	-1.80
31	1431	1448	-1.13
32	1366	1357	0.69
33	1076	1053	2.18
34	1165	1167	-0.19
35	1120	1138	-1.56
36	1263	1240	1.19
37	1219	1234	-1.18
38	1331	1329	0.13
39	1199	1183	1.33
40	1364	1393	-2.10
41	1411	1403	0.58
42	1428	1451	-1.60
43	1397	1409	-0.90
44	1435	1433	0.12
45	1436	1452	-1.09
46	1484	1461	1.57
47	1390	1376	1.03
48	1460	1455	0.32
49	1221	1239	-1.41
50	1139	1142	-0.26
51	1245	1233	1.01
52	1195	1213	-1.50
53	1273	1258	1.18
54	1188	1191	-0.18
55	1229	1245	-1.29
56	1179	1150	2.57
57	1489	1503	-0.89
58	1416	1388	2.04
59	1476	1459	1.23
60	1404	1411	-0.47
61	1459	1442	1.20
62	1394	1418	-1.71
63	1412	1428	-1.09
64	1349	1353	-0.27

Table 12.6.3: Estimated and simulated CO_2 concentration and the relative residuals.

12.6. A TERRESTRIAL BIOSPHERE METAMODEL

run	estimated CO_2 concentration ($\hat{Y}$)	simulated CO_2 concentration (Y)	relative residuals $100 * ((Y - \hat{Y})/Y)$
65	1469	1491	−1.51
66	1344	1328	1.19
67	1487	1476	0.77
68	1345	1376	−2.25
69	1484	1460	1.66
70	1482	1492	−0.66
71	1463	1470	−0.48
72	1457	1439	1.29
73	1277	1282	−0.41
74	1163	1145	1.59
75	1300	1282	1.42
76	1170	1175	−0.43
77	1301	1298	0.19
78	1282	1305	−1.78
79	1268	1289	−1.68
80	1245	1229	1.32
81	1507	1494	0.89
82	1503	1509	−0.42
83	1518	1520	−0.13
84	1518	1497	1.42
85	1339	1359	−1.45
86	1477	1472	0.34
87	1310	1308	0.16
88	1466	1480	−0.92
89	1260	1248	0.94
90	1296	1321	−1.88
91	1295	1320	−1.88
92	1334	1322	0.90
93	1158	1160	−0.22
94	1307	1283	1.82
95	1135	1113	1.97
96	1302	1318	−1.26
97	1449	1472	−1.58
98	1349	1346	0.22
99	1496	1484	0.82
100	1387	1405	−1.19
101	1473	1456	1.20
102	1461	1466	−0.36
103	1465	1477	−0.87
104	1444	1417	1.88
105	1295	1308	−0.99
106	1142	1115	2.42
107	1302	1280	1.73
108	1138	1148	−0.86
109	1296	1276	1.53
110	1274	1300	−1.98
111	1270	1292	−1.74
112	1239	1237	0.20
113	1431	1402	2.05
114	1437	1446	−0.61
115	1481	1499	−1.21
116	1499	1475	1.64
117	1339	1357	−1.36
118	1452	1446	0.44
119	1333	1319	1.07
120	1458	1486	−1.88
121	1251	1244	0.59
122	1279	1299	−1.61
123	1270	1286	−1.25
124	1308	1305	0.28
125	1113	1133	−1.78
126	1292	1266	2.07
127	1099	1077	2.13
128	1290	1292	−0.20

Table 12.6.3 (Continued).

extra run	Y	$\hat{Y}$	$100 * (Y - \hat{Y})/Y\,(in\%)$
129	1260.33	1268.48	−0.65
130	1128.98	1191.01	−5.49
131	1432.58	1484.19	−3.61
132	1217.65	1218.82	−0.10
133	1232.28	1248.99	−1.36
134	1248.88	1263.05	−1.13
135	1233.89	1244.53	−0.86
136	1216.85	1239.17	−1.83
137	1448.75	1474.09	−1.75
138	1372.28	1405.28	−2.40
139	1466.94	1495.61	−1.95
140	1422.69	1426.27	−0.25

Table 12.6.4: Extra runs.

Table 12.6.4 shows that for the twelve extra runs, all relative residuals are negative. This might be due to the fact that no random design is used for these extra runs.

12.6.3 Conclusions

The relative residuals appear to be less than 6%. Therefore the metamodel is accepted. The priority order of the factors in the terrestrial biosphere module influencing the atmospheric CO_2 concentration is, in decreasing order of importance:

x_3 : biotic growth factor;
x_{27} : humification factor of grassland;
x_{15} : residence time of humus in grassland;
x_{38} : shift from tropical forest to agricultural land;
x_{40} : life span of leaves in temperate forest;
x_5 : fraction of charcoal by burning of leaves, in temperate forest;
x_6 : fraction of charcoal by burning of leaves, in grassland;
x_{37} : shift from tropical forest to grassland;
x_{28} : humification factor of agricultural land;
x_{16} : residence time of humus in agricultural land;
x_{25} : humification factor of tropical forest;
x_{52} : life span of stems in temperate forest; and
x_{30} : humification factor of semi-desert area.

12.7 General Conclusions

Specification of a metamodel should be based on internal knowledge of the observed system. Since no statistic (such as Student's t) is available to support the validation process, doubt about the correctness of the metamodel has to be checked by examining a more complex metamodel.

To calibrate the regression model, many runs can be saved by choosing an experimental design, associating some factors with unused interactions between other factors. It turned out that 2^{k-p} designs are appropriate for estimating first-order effects and two-factor interactions, while central composite designs appeared to be suitable for estimating higher order effects, such as quadratic terms.

For the module of the costs of dike raising an adequate metamodel has been found, using a 2^{k-p} experimental design. Ignoring the dominant factor of economic growth, particularly the marginal costs of dike raising per meter are of major importance. The net effect of the Antarctic, the initial costs of dike raising, and the natural sea level rise play an equal, although less important role. The other factors, such as the minimal dike raising, melting of the glaciers, depth of ocean mixed layer, melting of the Greenland ice cap, and the thermal expansion delay time, are significant, but clearly of minor importance. Additionally one interaction, namely that between the marginal costs of dike raising per meter and the net effect of the Antarctic, is significant. The use of metamodels compels the user to examine the simulation model thoroughly. In this way errors or misuse of the simulation model can be located, as appeared when analyzing the dike raising model.

For the ocean module an adequate (quadratic) metamodel has been specified as well. The total thickness of the nine deep ocean layers largely influences the atmospheric CO_2 concentration. It is remarkable that the quadratic effect of this factor is also significant. Furthermore the thickness of the cold surface layer is of major importance. Less influential are the thickness of the warm ocean surface layer, the residence time of CO_2 in the cold surface layer and some interactions, for instance those between the thickness of the cold surface layer and the total thickness of the nine deep ocean layers.

The attempt to find a metamodel for the complicated terrestrial biosphere module also succeeded. This metamodel showed that the biotic growth factor, representing the fertilization effect, dominates the other factors. Amongst the other twelve significant factors especially the humification factor of grassland, the residence time of humus in grassland and the shift of land from tropical forest to agricultural land are of substantial importance. No significant interactions between factors have been found in this metamodel.

Summarizing it can be concluded that the techniques of metamodelling and experimental design can serve as a helpful tool in performing sensitivity analysis on simulation models. Even complex simulation models can be analyzed in this way, although the specification of adequate metamodels is very time-consuming. So far the metamodel for the terratrial biosphere is one of the largest known in the literature. This underlines the underestimation of metamodelling as a helpful tool in analyzing simulation models.

In the near future it is intended to specify metamodels for all modules of IMAGE. These metamodels will then be used for building an interactive version of IMAGE.

Chapter 13

Discussion

Over the last few years the usefulness of an integrated methodology to evaluate different future worlds with respect to climate change has been proved extensively. Demonstrations given for high level Dutch policy officials and in Parliament have proved the integrated greenhouse model IMAGE to be a helpful tool to improve the understanding of the greenhouse problem. Besides this, IMAGE has been used to increase awareness among all kinds of societal groups. IMAGE gives an overview of the complex problem by making explicit long-term relationships between causes and effects of the enhanced greenhouse effect, based on the latest scientific evidence, but without requiring detailed scientific knowledge from the user. With IMAGE it is possible to derive sets of allowable emissions from certain targets for climate impacts. It also enables scientists working in the field of climate change to put their efforts in a wider perspective. The model can easily be adapted to changed or increased knowledge.

In this study four scenario sets have been developed and calculated with IMAGE. Evaluating the results obtained from these scenario calculations leads to interesting results. For CO_2 these IMAGE simulations point out that continuation of the recent emissions trend of CO_2 leads to a rapid rise of the atmospheric CO_2 concentration, and may lead to a doubling of the CO_2 concentration around 2060. Even radical measures (scenario D: forced trends scenario) cannot prevent the CO_2 concentration from increasing, although in this case a CO_2 doubling is not reached before 2100. The role of deforestation within the carbon cycle has been investigated by the development of a separate deforestation module. It appeared that the primary direct cause of deforestation is the demand for agricultural land to satisfy demand for food, feed or debt-resolving export products. Simulation results with IMAGE show that if present deforestation rates continue a total destruction of the tropical forests will occur halfway through the next century. Although it is difficult to give quantified estimates, IMAGE simulations make clear that soil degradation is a process that, while seldom mentioned in this connection, contributes considerably to the rate of tropical deforestation by decreasing the availability of land for agriculture and pasture. Although reforestation can play a significant role in a transition period, a gradual shift towards renewables, away from fossil fuels, is a more effective strategy for mitigating climatic change. The contribution of biospheric changes to the greenhouse effect is important but small as compared to that of fossil fuel combustion. Model simulations showed a difference of a maximum of 10% in CO_2 concentrations over the next century between optimistic and pessimistic deforestation cases. The result, that even fast deforestation has only a moderate net effect on the carbon cycle in IMAGE, is primarily caused by an increasing soil carbon pool due to the conversion of organic matter into charcoal during burning, by CO_2 fertilization and by a possible underestimation of soil losses after conversion. To some extent the biospheric CO_2 emissions by deforestation will be counteracted by a CO_2 fertilization effect. If not for reasons of global climate change, deforestation should be stopped for other reasons, including the combating of erosion, the conservation of species diversity, the threat to the indigenous population and the effects on local and regional climate.

With the help of the developed simulation model for the CH_4-CO-OH cycle the importance of methane as a greenhouse gas is underlined. The model shows that there is no sense in forcing back only the methane emissions. Even if the emissions could be stabilized the methane concentrations would still rise, due to the interrelated influence of other trace gases such as carbon monoxide and hydroxyl (by NO_x and O_3). The sources of NO_x and CO_2 are especially important. Nevertheless the relative role of methane is expected to decrease because methane has a relatively short lifetime, the potential development of sources tends to be slower than that of other gases, and emission reduction and recovery measures are possible. However, should potentially important feedback mechanisms induced by wetlands and methane hydrates materialize, methane concentrations might increase considerably.

With respect to N_2O it is clear that in the long run N_2O will become a greenhouse gas of great importance. Even the most optimistic N_2O emission scenario leads to a rise in temperature. Policy measures can probably affect the N_2O concentrations, but additional research is needed to improve our knowledge of the sources and the validity of measurement techniques.

According to IMAGE calculations, the effect of the implementation of the Montreal Ozone Protocol will lead to a stabilization of the relative role of CFCs in the greenhouse effect, and is a significant step forward. However under the present Protocol conditions CFC concentrations can still increase. To decrease this role, it seems to be necessary to improve the agreement sharply.

Continuation of the recent trend of emissions of the relevant trace gases leads to a rapid increase of the mean global temperature. IMAGE simulations give an equilibrium temperature range of 2 to 8 degrees Celsius at the end of the next century. Calculations with IMAGE suggest that the rate of global temperature increase can be delayed considerably and limited to values which do not go beyond past climate experience by the global implementation of CO_2 emission reductions of 20% in 2000 and 50% in 2025. These figures are consistent with the recommendations of the Toronto conference on the Changing Atmosphere in 1988, provided these reductions are applied globally and a strong effort is undertaken to control the emissions of the other greenhouse gases, including a near phase-out of CFCs. Most effective are policies towards energy conservation and non-fossil fuels, which are presently unattractive given the low energy prices. Increase of efficiency with low inputs in the agricultural sector also contributes, albeit moderately, to the delay of the greenhouse effect.

Simulation experiments with IMAGE show for the end of the next century a sea level rise range of about 0.45 to 0.95 meter. The simulated increase for the next hundred years falls within the interval of 0.28 to 0.78 meter, which means an increase of about 2 to 5 times that of the past centuries. The dominant causes of this sea level rise will be the thermal expansion of the ocean, and the melting of glaciers and small ice caps. For the next hundred years changes in the mass of polar ice sheets may have a significant effect on sea level, but will be minor components compared to expansion and glaciers.

Simulations with IMAGE suggest that the socio-economic consequences of the greenhouse effect for the Netherlands are far-reaching, involving considerable cumulative costs for the next century. However, the yearly costs will be less than a half percent of the gross national product, indicating that, if policy makers will anticipate in time, the socio-economic costs can be mitigated.

In order to evaluate recent future proposals for the control of greenhouse gas emissions it is recommended to determine long-term environmental or climate goals that would allow for a sustainable development. Both the Temperature Increasing Potential (TIP) and the Sea level Rise Potential (SRP) can serve as useful tools for setting environmental long-term goals with respect to climate change. The TIP and SRP provide an index to compare the temperature increasing effect and the sea level rise of greenhouse gas emissions, as a greenhouse counterpart to the ozone depleting potential (ODP). These TIPs and SRPs can be used to define quantified environmental targets which can serve as reference values for the development of international response strategies. Although surrounded by many uncertainties it is possible to estimate such indexes with IMAGE. However simulations demonstrate the dynamical element in determining the TIP. Hence it is better to speak of a TIP function rather than a TIP value.

Results based on calculations with IMAGE showed that CH_4 and N_2O in particular, being much more effective than CO_2, are currently underestimated. This appeared from the relative contributions of greenhouse gases to the global warming for the year 1985, calculated according to the TIP concept, which showed a considerable share of especially CH_4. Based on the TIP estimates it can be concluded that CH_4 and N_2O will have the most threatening greenhouse potential for the future.

The technique of metamodelling might be extremely useful in long-term scenario studies. Up to now this technique has been underestimated as a powerful tool for analyzing sensitivities and uncertainties in simulation models. For different scientific topics, ranging from environmental protection, health care to economics, scenario studies on future developments are marked by high degrees of uncertainties. Living in a world with limited financial means, policy makers have to make choices in initiating projects to reduce these uncertainties. Sensitivity analysis by metamodelling and experimental design deserves more attention, and can provide a method for differentiating between the relative relevance of these uncertainties and can therefore play a decisive role in determining priorities for filling gaps in our scientific knowledge.

References

Abrahamse, A.H., Baarnse, G., Beek, E. van (1982a), *"Policy analysis of water management for the Netherlands"*, Volume 11, Rijkswaterstaat, Government printing office, The Hague, The Netherlands.

Abrahamse, A.H., Baarnse, G., Beek, E. van (1982b), *"Policy analysis of water management for the Netherlands"*, Volume 12, Rijkswaterstaat, Government printing office, The Hague, The Netherlands.

Abrahamse, A.H., Baarnse, G., Beek, E. van (1982c), *"Policy analysis of water management for the Netherlands"*, Volume 16, Rijkswaterstaat, Government printing office, The Hague, The Netherlands.

AGGG (1990), *"Targets and indicators of climatic change"*, Report of Working Group 2 of the Advisory Group on Greenhouse Gases (AGGG).

Aldenberg, T.A. (1988), *"Note concerning integrated modelling"* (in Dutch), RIVM, Bilthoven, The Netherlands.

Anderson, I.C., and Levine, J.S. (1987), *"Simultaneous field measurements of biogenic emissions of nitric oxide and nitrous oxide"*, Journal of Geophysical Research 92, 965–976.

Atmospheric Ozone (1985), *"Assessment of our understanding of the process controlling its present distribution and change"*, volume 1; WMO/ NASA/UNEP/FVA/NOAA/CEC.

Augustsson, T., Ramanathan, V (1977), *"A radiative-convective model study of the CO_2 climate problem"*, Journal of the Atmospheric Sciences 34, 448–451.

Automobile International (1929–1987) World Automotive Market.

Avenhaus, R. and Hartmann, G. (1975), *"The carbon cycle of the earth - a material balance approach"*, Research Report no. RR-75-45, IIASA, Laxenburg, Austria.

Banin, A. (1986), *"Global budget of N_2O : the role of soils and their change"*, The Science of the Total Environment 55, 27–38.

Barnett, T.P. (1983), *"Recent changes in sea level and their possible causes"*, Climatic Change 5, 15–38.

Barnett, T.P. (1985), in Detecting the climatic effect of increasing carbon dioxide, McCracken, M.C. and Luther, F.M., (eds), US Department of Energy, Washington D.C., 91–107.

Barth, M.C., and Titus, J.G. (1984), *"Greenhouse effect and sea level rise: a challenge for this generation"*, van Nostrand Rheinhold Company Inc., New York.

Bartlett, K.B., Harris, R.C. and Seiler, D.I. (1985), *"Methane flux from coastal salt marshes"*, Journal of Geophysical Research 90, 5710–5720.

Beafort, G.A., Baarse, G., Pluym, M., Roelse, P., and Peerbolte, E.B. (1989), *"Beach and dune depletion"* (in Dutch), Coastal defence after 1990, Technical Report 11, The Netherlands.

Bettonvil, B. and Kleijnen J.P.C. (1988), *"Measurement scales and resolution IV designs: a note"*, to be published in the Americain Journal of Mathematical and Management Sciences, Catholic University of Brabant, Tilburg, the Netherlands.

Bettonvil, B. (1990), *"Detection of important factors by sequential bifurcation"*, Thesis, Catholic University of Brabant, Tilburg, The Netherlands.

Bettonvil, B. and Rotmans, J. (1990), *"Screening designs, applied to the greenhouse simulation model IMAGE"*, to be published.

Bingemer, H.G. and Crutzen, P.J. (1987), *"The production of methane from solid wastes"*, Journal of Geophysical Research 92, 2181–2187.

Björkström, A. (1979), *"A model of CO_2 interaction between atmosphere, oceans, and land biota"*, in The Global Carbon Cycle, B. Bolin, E. Degens, S. Kempe, and P. Ketner (eds.), SCOPE 13, John Wiley and Sons, New York.

Bolin, B., Björkström, A., Holmen, K., and Moore, B. (1983), *"The simultaneous use of tracers for ocean circulation studies"*, Tellus 35B, 206–236.

Bolin, B., Döös, B.R., Jäger, J., and Warrick, R.A. (eds) (1986), *"The greenhouse effect, climatic change and ecosystems"*, SCOPE 29, Chichester, UK, John Wiley & Sons.

Bolin, B. (1986), *"How much CO_2 will remain in the atmosphere: the carbon cycle and projections for the future"*, In: 'The Greenhouse Effect, Climatic Change and Ecosystems', B.Bolin, B.R. Döös, J.Jäger, R.A.Warrick (eds), SCOPE 29, Chichester, UK, John Wiley and Sons, 93–155.

Bolle, H.J., Seiler,W. and Bolin, B. (1986), *"Other greenhouse gases and aerosols: assessing their role for atmospheric radiative transfer"*. In: The Greenhouse Effect, Climatic Change, and Ecosystems, B. Bolin, B.R. Döös, J. Jäger and R.A. Warrick (eds.), SCOPE 29, Chichester, UK, John Wiley and Sons, 157–203.

Boois, H. de, Rotmans, J., Swart, R.J. (1988), *"State of affairs and perspectives of the greenhouse problem: presentation of IMAGE for the Dutch parliament"* (in Dutch), Report no. 758471005, RIVM, Bilthoven, the Netherlands.

Bouwman, A.F. (1989a), *"Land evaluation for dry farming"*, working paper, International Soil Reference and Information Centre, Wageningen.

Bouwman, A.F. (1989b), *The role of soils and land use in the 'greenhouse effect'*, Accepted by the Netherlands Journal of Agricultural Science.

Box, G.E.P., Hunter, W.G., and Hunter, J.S. (1978), *"Statistics for experimenters"*, J. Wiley & Sons Inc., New York.

Brasseur, G. and Rudder, A. de (1987), *"The potential impact on atmospheric ozone and temperature of increasing trace gas concentrations"*, Journal of Geophysical Research 92, 10903–10920.

Brewer, P.G. (1983), *"Carbon dioxide and the oceans"*, In: Changing Climate, Report of the Carbon Dioxide Assessment Committee, NRC/NAS/NAE/IOM, National Academy Press, Washington D.C.

Brouwer, F.M. (1987), *"Integrated environmental modelling: design and tools"*, Kluwer Academic Publisher Group, Dordrecht-Boston- Lancaster.

Bruggeman, G.A. (1988a), *"Notice concerning the salt intrusion in the New Waterway"*, RIVM, Bilthoven.

Bruggeman, G.A. (1988b), *"Increase in salt seepage along the Dutch coast with a sea level rise (global approach)"*, RIVM, Bilthoven.

Brühl, Ch., and Crutzen, P.J. (1985), *"A time dependent 1-D photochemical climate model"*, Progress Report to contract CLT-080-D.

Brühl, Ch. and Crutzen, P.J. (1988), *"Scenarios of possible changes in atmospheric temperatures and ozone concentrations due to man's activities, estimated with a one-dimensional coupled photochemical model"*, Climate Dynamics 2, 173–203.

Buringh, P., Heemst, H.D.J. van, and Staring, G.J. (1975), *"Computation of the maximum food production of the world"*, Agricultural University, Wageningen.

Burke, M.K., Houghton, R.A. and Woodwell, G.M. (1990), *"Increased emissions of CH_4 from wetlands as a consequence of global warming"*, in Bouwman, A.F. (ed.): Soils and the Greenhouse Effect, Wiley, London, 1990.

Central Planning Bureau (1984), *"Dutch economy in the long-term, three scenarios for the period 1985-2010"* (in Dutch), The Netherlands.

Cess, R.D., and Goldenberg, S.D. (1981), *"The effect of ocean heat capacity upon global warming due to increasing atmospheric carbon dioxide"*, Journal of Geophysical Research 86, 498–502.

Chamberlain, J.W., Foley, H.M., MacDonald, G.J., and Ruderman, M.A. (1982) *"Climatic effects of minor atmospheric constituents"*, in Carbon Dioxide Review, W.C. Clark (ed), Oxford University Press, New York, 255–295.

Chemical Manufacturers Association (CMA) (1983), *"Production, sales and calculated release of CFC-11 and CFC-12 through 1982"*, News Release of the CMA, Washington D.C.

Chemical Manufacturers Association (CMA) (1987), *"Production, sales and calculated release of CFC-11 and CFC-12 through 1986"*, News Release of the CMA, 18 November 1987, Washington DC.

Cheng, H.C., Steinberg, M. and Beller, M. (1986), *"Effects of energy technology on global CO_2 emissions"*, US Department of Energy, Report no. TR030, Washington, D.C.

Conrad, R., and Seiler, W. (1980), *"Field measurements of the loss of fertilizer nitrogen into the atmosphere as nitrous oxide"*, Atmospheric Environment 14, 555–558.

Conrad, R., Seiler, W., and Bunse, G. (1983), *"Factors influencing the loss of fertilizer nitrogen into the atmosphere as N_2O "*, Journal of Geophysical Research 88, 6709–6718.

Crutzen, P.J. (1983), *"Atmospheric interactions - homogeneous gas reactions of C, N, and S containing compounds"*, in The Major Biogeochemical Cycles and their Interactions, B. Bolin and R.B. Cook (eds), SCOPE 21, John Wiley and Sons, New York, 67–112.

Crutzen, P.J. (1985), *"The role of the tropics in atmospheric chemistry"*, in Geophysiology of Amazonia, R. Dickinson (ed.), New York.

Crutzen, P.J., Asselmann, I. and Seiler, W. (1986a), *"Methane production by domestic animals, wild ruminants, other herbivorous fauna and humans"*, Tellus 38B, 271–284.

Crutzen, P.J. and Graedel, T.J. (1986b), *"The role of atmospheric chemistry in environment-development interactions"*. In: 'Sustainable Development of the Biosphere', W.C.Clark and R.E.Munn (eds), IIASA, Laxenburg.

Cunnold, D.M., Prinn, R.G., Rasmussen, R.A., Simmonds, P.G., Alyea, F.N., Cardelino, C.A., Crawford, A.J., Fraser, P.J., and Rosen, R.D. (1983a), *"The atmospheric lifetime experiment 3. Lifetime methodology and application to three years of $CFCl_3$ data"*, Journal of Geophysical Research 88, 8379–8400.

Cunnold, D, M., Prinn, R.G., Rasmussen, R.A., Simmonds, P.G., Alyea, F.N., Cardelino, C.A., Crawford, A.J., Fraser, P.J., and Rosen, R.D. (1983b), *"The atmospheric lifetime experiment 4. The results for CF_2Cl_2 based on three years of data"*, Journal of Geophysical Research 88, 8401–8414.

Dantzig, D. van (1956), *"Economic decision problems for flood prevention"*, Econometrica 24, 276–287.

Delft Hydraulics (1990), *"Sea level rise: a world wide cost estimate of basic coastal defence measures"*, draft Report for IPCC/Rijkswaterstaat.

Deltacommissie (1960), *"Eindverslag en interimadviezen"*, deel 1, pp. 56 etc., Government printing office, The Hague.

Derwent, R.G. (1990), *"Trace gases and their relative contribution to the greenhouse effect"*, Modelling and Assessments Group, Environmental and Medical Sciences Division, Harwell Laboratory, Oxfordshire, United Kingdom.

Detwiler, R.P. and Hall, C.A.S. (1985), *"Land use change and carbon exchange in the tropics: II. estimates for the entire region"*, Environmental Management 9, no. 4, 335–344.

Detwiler, R.P. and Hall, C.A.S. (1988), *"Tropical forests and the global carbon cycle"*, Science 239, 42–47.

Dickinson, R.E. (1986), *"How will climate change? The climate system and modelling of future climate"*. In: 'The greenhouse effect, climatic change, and ecosystems', B. Bolin, B.R. Döös, J. Jäger, and R.A. Warrick (eds), SCOPE 29, Chichester, UK, John Wiley and Sons, 206–270.

Dillingh, D., Graaf, J. van de, Vellinga, P., Visser, C. (1984), *"Leidraad voor de beoordeling van de veiligheid van duinen als waterkering"*, Technische Adviescommissie voor de Waterkering, Staatsdrukkerij, The Hague, The Netherlands.

Dover, M. and Talbot, L.M. (1987), *"To feed the earth: Agro-ecology for sustainable development"*, World Resources Institute, Washington D.C.

Dubbeld, M. (1985), *"De invloed van het bewonersgedrag, de bouwkundige en installatietechnische varianten en de regio op het energiegebruik van een eengezinswoning, berekend met een meerkamermodel"*, IMG-TNO, Delft, The Netherlands.

Edmonds, J.A., and Reilly, J.M. (1983), *"A long-term global energy-economic model of carbon dioxide release from fossil fuel use"*, Energy Economics, April 1983, 283–297.

Edmonds, J.A. and Reilly, J.M. (1985), *"Future global energy and carbon dioxide emissions"*. In: 'Atmospheric carbon dioxide and the global carbon cycle', US Department of Energy, Washington D.C.

Edmonds, J.A. and Reilly, J.M. (1986), *"The long-term global energy CO_2 model: PC-version A84PC"*; Carbon Dioxide Information Center, Oak Ridge.

Efron, B. and Gong, G. (1983), *"A leisurely look at the bootstrap, the jacknife and cross-validation"*, American Statistician 37, 36–48.

Ehhalt, D.H. (1985), *"Methane in the global atmosphere"*, Environment 27 (no. 10), 6–11.

Elshout, A.J. (1989), *"N_2O in the atmosphere"*, Report of the divison Research and Development of the KEMA no. 80244-MOA 89-3250, Arnhem, the Netherlands (in Dutch).

Elzen, M.G.J. den, and Rotmans, J. (1988), *"Simulationmodel for the socio-economic consequences of the greenhouse effect for the Netherlands"*, (in Dutch), Report no. 758471008, RIVM, Bilthoven, The Netherlands.

Elzen, M.G.J. den, and Rotmans, J. (1989), *"A scenario study on the socio-economic impact of the greenhouse effect for the Netherlands"*, Report M 89-09, University of Limburg.

Elzen, M.G.J. den, and Rotmans, J. (1990), *"A scenario study on the socio-economic consequences of a sea level rise for the Netherlands"*, accepted by Climatic Change.

Elzen, M.G.J. den, Rotmans J., and Vrieze, O.J. (1990a), *"Mathematical models for strategic dike raising"*, in press.

Elzen, M.G.J. den, Rotmans, J., and Swart. R.J. (1990b), *"The role of CFCs and substitutes and other halogenated chemicals in climate change"*, Report, RIVM, Bilthoven, the Netherlands, in press.

Elzen M.G.J. den, Rotmans, J., and Swart, R.J. (1990c), *"The greenhouse effect of halocarbons after the Montreal Protocol"*, submitted to International Environmental Affairs.

Emanuel, W.R., Killough, G.G., Stevenson, M.P., Post, W.M., and Shugart, H.H. (1984), *"Computer implementation of a globally averaged model of the world carbon cycle"*, DOE/NBB-0062, U.S. Department of Energy, Washington, D.C.

Emanuel, W.R., Fung, I.Y.S., Killough, G.G., Moore, B., and Peng, T.H. (1985), *"Modelling the global carbon cycle and changes in the atmospheric carbon dioxide levels"*, in 'Atmospheric carbon dioxide and the global carbon cycle', J.R. Trabalka (ed.), United States Department of Energy, DOE/ER/-0239-1, Washington D.C.

Environment Canada (1988), *"Conference statement: the changing atmosphere: implications for global security"*, Toronto, Canada.

Environmental Protection Agency (1989), *"Policy options for stabilizing global climate"*, Draft Report to Congress, Washington, D.C.

EPA (1989), *"Policy options for stabilizing global climate"*, Draft Report to Congress, D.A. Lashof and D.A. Tirpack (eds.), United States Environmental Protection Agency, Office of Policy, Planning and Evaluation, Washington, D.C.

EPA (1989), *"Policy options for stabilizing global climate"*, draft Report to Congress, Washington D.C., 1989.

ESRG (1988) LEAP, *"LDC energy alternatives planning system"*, User's Guide, Energy Systems Research Group and the Beijer Institute, Boston.

Esser, G. (1987), *"Sensitivity of global carbon pools and fluxes to human and potential climatic impacts"*, Tellus 39B, 245–260.

Eversdijk, P.J. (1989), *"Hard coastal defence, sea dikes, harbour areas and beach walls as dams"*, (in Dutch), Coastal defence after 1990, Technical Report 16, Rijkswaterstaat, The Netherlands.

Fearnside, P.M. (1987), *"Causes of deforestation in the brazilian amazon"*, in: Dickenson, R.E. (ed.): 'The Geophysiology of Amazonia', Wiley.

Fisher, C.R., Hales, C., Wang, W.C., Ko, M., and Sze, N. (1990), *"Model calculations of the relative effects of CFCs and their replacements on global warming"*, Nature 344, 513–516.

Food and Agricultural Organization (FAO) (1984), *"Land, food and people"*, Rome, 1984.

Food and Agricultural Organization (FAO) (1986), *"Tropical forest action plan"*, Rome.

Food and Agricultural Organization (FAO) (1987), *"Agriculture: towards 2000"*, Rome.

Gamlen, P.H., Lane, B.C., Midgley, P.M., and Steed, J.M. (1986), *"The production and release to the atmosphere of CCl_3F and CCl_2F_2 (Chlorofluorocarbons CFC-11 and CFC-12)"*, Atmospheric Environment 20, 1077–1085.

GEMS Monitory and Assessment Research Center (1989), United Nations Environment Program, Environmental Data Report, London, UK, in cooperation with the World Resource Institute (WRI), and UK Department of the Environment.

Gardner, R.H., Rojder, B., Bergström, U. (1983), *"PRISM: a systematic method for determining the effect of parameter uncertainties on model predictions"*, Studsvik Energiteknik AB, Report Studsvik/NW-83/555, Nykoping, Sweden.

Goodman, M.R. (1974), *"Study note in system dynamics"*, Wright-Allen Press inc., Cambridge, Massachussets.

German Enquete Commission (1988), *"Zur Sache; schutz der Erdatmosphäre; eine internationale Herausforderung"*, German Bundestag.

Gibbs, M.J., Inglis, M.R., Linder, K.P. (1987), *"Potential impacts of climate change on electric utilities"*, New York State Energy Research and Development Authority, New York.

Goldenberg J., Johansson T.B., Reddy A.K.N., Williams R.H. (1987a), *"Energy for development"*, World Resources Institute, Washington D.C.

Goldenberg, J., Johansson, T.B., Reddy, A.K.N. and Williams, R.H. (1987b), *"Energy for a sustainable world"*, World Resources Institute, Washington, D.C.

Gornitz, V., Lebedeff, L., and Hansen, J. (1982), *"Global sea level trend in the past century"*, Science 215, 1611–1614.

Gornitz, V. et. al. (1984), *"Global sea level trend in the past century"*, Science 226, 1418–1421.

Goudriaan, J. (1987), *"The biosphere as driving force in the global carbon cycle"*, Netherlands Journal of Agricultural Science 35, 177–187.

Goudriaan, J. (1988a), *"Agriculture and climate"*, Lucht en omgeving, March.

Goudriaan, J. (1988b), *"modelling biospheric control of carbon fluxes between atmosphere, ocean and land in view of climatic change"*, draft, to be published, Wageningen, The Netherlands.

Goudriaan, J. and Ketner, P. (1984), *"A simulation study for the global carbon cycle including man's impact on the biosphere"*, Climatic Change 6, 167–192.

Haan, B. de: (1989), *"Note concerning ocean modelling"*, RIVM, Bilthoven, The Netherlands (in Dutch).

Hafkamp, W.A. (1984), *"Economic-environmental modelling in a national-regional system"*, North Holland Publishing Company, Amsterdam, The Netherlands.

Hageman, H. (1988), Personal communication, Rijkswaterstaat, The Netherlands.

Hahn, J. (1974), *"The North Atlantic ocean as a source of atmospheric N_2O"*, Tellus 26, 160–168.

Hahn, J. and Junge, C. (1977), *"Atmospheric nitrous oxide: A critical review"*, Zeitschrift Naturforschung 32, 190–214.

Ham, G. van, Rotmans, J., and Kleijnen, J.P.C. (1990), *"Sensitivity analysis by metamodels and experimental designs, applied to the greenhouse simulation model IMAGE"*, submitted to Kwantitatieve Methoden, Heiloo, the Netherlands.

Ham, J. van, (1987), *"Tropospheric chemistry: proposal for a dutch research program in an international framework"* (in Dutch), The Council for Environment and Nature Research, The Hague, the Netherlands.

Hameed, S., Pinto, J.P. and Stewart, R.W. (1979), *"Sensitivity of the predicted CO-OH-CH_4 perturbation to tropospheric NO_x concentration"*, Journal of Geophysical Research 84 (C2), 763–768.

Hammitt, J.K. et. al. (1986), *"Product uses and market trends for potential ozone-depleting substances 1985-2000"*, Rand Corporation.

Hansen, J., Russel, G., Lacis, A., Fung, I., Rind, D., and Stone, P. (1985), *"Climate response times: dependence on climate sensitivity and ocean mixing"*, Science 229, 857–859.

Hansen, J., Fung, I., Lacis, A., Rind, D., Lebedeff, S., Ruedy, R., Russel, G., and Stone, P. (1988), *"Global climate changes as forecast by Goddard Institute for Space Studies three-dimensional model"*, Journal of Geophysical Research 93, 9341–9364.

Hao, W.M., Wofsy, S.C., McElroy, M.B., Beer, J.M., and Toqan, M.A. (1987), *"Sources of atmospheric nitrous oxide from combustion"*, Journal of Geophysical Research 92, 3098–3104.

Harvey, L.D. Danny (1989a), *"Transient climatic response to an increase of greenhouse gases"*, Climatic Change 15, 15–30.

Harvey, L.D. Danny (1989b), *"Managing atmospheric CO_2"*, Climatic Change 15, 343–381.

Hasselmann, K. (1988), *"Climate and development: scientific efforts and assessment the state of the art"*, paper presented at the World

Congress, Climate and Development: Climatic Change and Variability and the Resulting Social, Economic and Technological Implications, Hamburg, FRG, 7–10 November, 1988.

Health Council of the Netherlands (1983), *"First advice concerning the CO_2 problem"* (in Dutch), Staatsuitgeverij, The Hague, The Netherlands.

Health Council of the Netherlands (1986), *"Second advice concerning the CO-problem: scientific insights and social consequnces"* (in Dutch), Staatsuitgeverij, The Hague, The Netherlands.

Hecht, S. (1988), *"Livestock expansion in the Brazilian tropics: dynamics and consequences"*, paper presented at the 46th International Congress of Americanists, Amsterdam.

Henderson-Sellers, A., and McGuffie, K. (1987), *"A Climate Modelling Primer"*, Research and developments in climate and climatology, A. Henderson-Sellers and M.M. Verstraete (eds.), John Wiley & Sons, Chichester.

Hettelingh, J.P. (1989), *"Uncertainty in modelling regional environmental systems: the generalization of a watershed acidification model for predicting broad scale effects"*, Thesis, Free University of Amsterdam, The Netherlands.

Hewitt, C.N. and Harrison, R.M. (1985), *"Tropospheric concentrations of the hydroxyl radical - a review"*, Atmospheric Environment 19, 545–554.

Hignett, T.P. (1985), *"Outlook for the fertilizer industry 1978-2000"*. In: Fertilizer Handbook, International Fertilizer Development Center, Muscle Shoals, Alabama.

Hill, R.D., Rinker, R.G., Coucouvinos, A. (1984), *"Nitrous oxide production by lightning"*, Journal of Geophysical Research 89, 1411.

Hoffert, M.I., Callegari, A.J., and Hsieh, C.T. (1981), *"A box-diffusion carbon cycle model with upwelling, polar bottom water formation and a marine biosphere"*, In: Carbon cycle modelling, SCOPE 16, B. Bolin (ed.), John Wiley and Sons, New York.

Hoffman, J.S., Keynes, D., and Titus, J.G. (1983), *"Projecting future sea level rise"*, Washington, D.C.: Government Printing Office.

Hoffman, J.S., Wells, J.B., and Titus, J.G. (1983), *"Projecting future sea level rise: methodology, estimates to the year 2100, and research needs"*, US GPO No. 055-000-00236-3, Government Printing Office, Washington D.C.

Hoffman, J.S. (1984), *"Estimates of future sea level rise"*, in: Greenhouse Effect and Sea Level Rise: a Challenge for this Generation, M.C. Barth, and J.G. Titus (eds.), van Nostrand Rheinhold Company Inc., New York.

Holzapfel-Pschorn, A. and Seiler, W. (1986), *"Methane emission during a cultivation period from an Italian rice paddy"*, Journal of Geophysical Research 92, 11803–11814.

Houghton, R.A., Hobbie, J.E., Mellillo, J.M., Moore, B. Peterson, B.J., Shaver, G.R. and Woodwell, G.M. (1983), *"Changes in the carbon content of terrestrial biota and soils between 1860 and 1980: a net release of CO_2 to the atmosphere"*, Ecological Monographs 53, no. 3, 235–262.

Houghton, R.A., Boone, R.D., Mellillo, J.M., Palm, C.A., Woodwell, G.M., Myers, N., Moore, B. and Skole, D.L. (1985), *"Net flux of carbon dioxide from tropical forests in 1980"*, Nature 316, 617–620.

Houghton, R.A., Boone, R.D., Fruci, J.R., Hobbie, J.E., Mellillo, J.M., Palm, C.A., Peterson, B.J., Shaver, G.R. and Woodwell, G.M. (1987), *"The flux of carbon from terrestial ecosystems to the atmosphere in 1980 due to changes in land use: geographic distribution of the global flux"*, Tellus 39B, 122–139.

Huizinga, P. (1988), Personal communication, Nederlandse Gasunie, Groningen, The Netherlands.

IEA (1989), *"World energy outlook"*, Draft Paper, IEA, Paris.

IPCC (1989), *"Emissions scenarios of the response strategies working group of the intergovernmental panel on climate change"*, draft Report of the US-Netherlands Expert Group on Emissions Scenarios, Bilthoven, the Netherlands, 7–8 april.

IPCC (1990), *"Scientific assessment of climate change"*, Report prepared for IPCC by working Group 1, pre-publication copy, June 1990.

Isaksen, I.S.A. and Høv, (1987), *"Calculation in trends in the tropospheric concentration of O_3, OH, CO, CH_4 and NO_x "*, Tellus 39B, 271–285.

Jackman, C.H., and Guthrie, P.D. (1985), *"Sensitivity of N_2O, $CFCl_3$ and CF_2Cl_2 ; two-dimensional distributions to O_2 absorption cross sections"*, Journal of Geophysical Research 90, 3919–3923.

Jäger, J. (1988), *"Developing policies for responding to climatic change"*. A Summary of the Discussions and Recommendations of the Workshops held in Villach (28 sept–2oct 1987) and Bellagio (9–13 nov 1987), under the Auspices of the Beijer Institute, Stockholm World Climate Program Impact Studies, WMO/UNEP, WMO/TD-No. 225, 53 pp.

Jansen, T. (1987), *"The terrestrial biospheric carbon cycle and the effect on the atmospheric CO_2 concentration"*, Report of the Energy Study Centre, Petten, The Netherlands (in Dutch).

Jansen, P.H.M., Slob, W., Rotmans, J. (1990), *"Uncertainty analysis and sensitivity analysis: an inventory of ideas, methods, and techniques from the literature"*, Report (in Dutch), forthcoming, Bilthoven, The Netherlands.

Jong, C.H. de (1985), *"De gewenste veiligheid tegen overstromingen vanuit economisch oogpunt"*, Rijkswaterstaat.

Jong, C.H. de (1986), *"De veiligheid van Nederland bij grote zeespiegelrijzing (een verkennend onderzoek)"*, Rijkswaterstaat.

Kavanaugh, M. (1987), *"Estimates of future CO, N_2O and NO_x emissions from energy combustion"*, Atmospheric Environment 21, 463–468.

Keeling, C.D., Bacastow, R.B., and Whorf, T.P. (1982), *"Measurements of the concentrations of carbon dioxide at Mauna Loa observatory, Hawaii"*, in Carbon Dioxide Review: 1982, 377–385, W.C. Clark (ed.), Oxford University Press, Oxford, United Kingdom.

Keller, M., Kaplan, W.A. and Wofsy, S.C. (1986), *"Emissions of N_2O, CH_4 and CO_2 from tropical forest soils"*, Journal of Geophysical Research 91, 11791–11802.

Keller, M., Kaplan, W.A. and Wofsy, S.C. (1987), *"Emissions of N_2O, CH_4 and CO_2 from tropical forest soils"*, Journal of Geophysical Research 91, 11791–11802.

Kellog, W.W. (1983), *"Feedback mechanisms in the climate system affecting future levels of carbon dioxide"*, Journal of Geophysical Research 88, 1263–1269.

Kerr, R.A. (1984), *"Doubling of methane supported"*, Science 226, 954–955.

Khalil, M.A.K. and Rasmussen, R.A. (1982), *"Secular trends of atmospheric methane (CH_4)"*, Chemosphere 11, no. 9, 877–883.

Khalil, M.A.K., and Rasmussen, R.A. (1983), *"Increase and seasonal cycles of nitrous oxides in the earth's atmosphere"*, Tellus 35B, 161–169.

Khalil, M.A.K. and Rasmussen, R.A. (1984), *"Carbon monoxide in the earth's atmosphere: increasing trend"*, Science 224, 55–56.

Khalil, M.A.K. and Rasmussen, R.A. (1985), *"Causes of increasing atmospheric methane: depletion of hydroxyl radicals and the rise of emissions"*, Atmospheric Environment 19, no. 3, 397–407.

Khalil, M.A.K. and Rasmussen, R.A. (1987), *"Atmospheric methane: trends over the last 10,000 year"*, Atmospheric Environment 21, 2445–2452.

Kiehl, J.T. and Dickinson, R.E. (1987), *"A study of the radiative effects of enhanced atmospheric CO_2 and CH_4 on early earth surface temperatures"*, Journal of Geophysical Research 92, 2991–2998.

Kleijnen, J.P.C. (1987), *"Statistical tools for simulation practitioners"*, Marcel Dekker, Inc., New York.

Kleijnen, J.P.C. (1990), *"Statistics and deterministic simulation models: why not?"*, Catholic University of Brabant, Report no. FEW 435, Tilburg, The Netherlands.

Kleijnen, J.P.C., Ham, G. van, and Rotmans, J. (1990), *"Techniques for sensitivity analysis of simulation models: a case study for the greenhouse effect"*, submitted to Simulation.

Knoester, D., Baarse, G., Kuik, A.J., Louisse, C.J. (1989), *"Systematical model"* (in Dutch), Coastal defence after 1990, Technical Report 17, The Netherlands.

Kohlmaier, G.H., Kratz, G., Bröhl, H., and Siré, E.O. (1981), *"The source-sink function of the terrestrial biota within the global carbon cycle"*, in Energy and Ecological modelling, W.J. Mitsch, R.W. Bosserman, and J.M. Klopatek (eds.), Elsevier Science Publ. Co., Amsterdam, 57–68.

Kohlmaier, G.H., Janecek, A., and Kindermann, J. (1990), *"Positive and negative feedback loops within the vegetation/soil system in response to a CO_2 greenhouse warming"*, in Soils and the greenhouse effect, Bouwman, A.F. (ed.), John Wiley and Sons, Chichester.

Kram, T. and Okken, P.A. (1989), *"Integrated assessment of energy-options for CO_2 reduction"*, in Climate and Energy: the Feasibility of Controlling CO_2 Emissions, P.A. Okken, R.J. Swart, and S. Zwerver (eds.), Kluwer Academic Publishers, Dordrecht, The Netherlands.

Kvenvolden, K.A. (1988), *"Methane hydrates and global climate"*, Geobiochemical Cycles, vol. 2, no. 3, 221–229.

Lammerts, I. (1989). *"Uncertainty analysis applied to the greenhouse model IMAGE"* (in Dutch) University of Utrecht, the Netherlands.

Langeweg, F. (ed.) (1989), *"Concern for tomorrow: a national environmental survey 1985–2010"*, Bilthoven, The Netherlands.

Lanly, J.P. (1982), *"Tropical forest resources"*, FAO, Rome.

Lashof, D.A. (1989), *"The dynamic greenhouse: feedback processes that may influence future concentrations of atmospheric trace gases and climatic change"*, to be published in Climatic Change.

Lashof, D.A., and Ahuja, D.R. (1990), *"Relative global warming potentials of greenhouse gas emissions"*, Nature, 344, 529–531, Washington, D.C.

Lashof, D.A., and Rotmans, J. (1990), *"Evaluation of methods for estimating Global Warming Potentials (GWPs) of greenhouse gases"*, submitted to Climatic Change.

Levine, J.S, Shaw, E.F. (1983), *"In situ aircraft measurements of enhanced levels of N_2O associated with thunderstorm lightning"*, Nature 303, 312.

Levine, J.S., Rinsland, G.P. and Tennille, G.M. (1985), *"The photochemistry of methane and carbon monoxide in the troposphere in 1950 and 1985"*, Nature 318, 254–257.

Logan, J.A., Prather, M.J., Wofsy, S.C. and McElroy, M.B. (1981), *"Tropospheric chemistry: a global perspective"*, Journal of Geophysical Research 86, 5163–5171.

Lowe, D.C., Brenninkmeijer, C.A.M., Manning, M.R., Sparks, R. and Wallaca, G. (1988), *"Radiocarbon determination of atmospheric methane at Baring Head, New Zealand"*, Nature 332, 522–524.

Lugo, A.E., and Brown, S. (1980), *"Tropical forest ecosystems: sources or sinks of atmospheric carbon"*, Unasylva 32 (129), 8–13.

Lugo, A.E., and Brown, S. (1986), *"Steady state terrestrial ecosystems and the global carbon cycle"*, Vegetatio 68, 83–90.

Lyon, R.K., Kramlich, J.C., and Cole, J.A. (1989), *"Nitrous oxide: sources, sampling and science policy"*, Environmental Science and Technology 23, (4), 1989.

Marland, G., and Rotty, R.M. (1984), *"Carbon dioxide emissions from fossil fuels: a procedure for estimation and results for 1950–1982"*, Tellus 36B, 232–261.

Marland, G. and Rotty, R.M. (1985), *"Greenhouse gases in the atmosphere: what do we know?"*, Journal of the Air Pollution Control Association 35, 1033–1038.

McKinsey & Company, Inc. (1989), *"Background paper on funding mechanisms"*, for the Ministrial Conference on Atmospheric Pollution and Climatic Change, with particular attention to Global Warming 6th and 7th november 1989.

Meadows, D.L., Randers, J., and Behrens, W. (1972), *"The limits to growth"*, Potomac Associates/Universe Books, New York.

Meier, M.F. (1984), *"Contribution of small glaciers to sea level"*, Science 226, 1418–1421.

Michael, P., Hoffert, M., Tobias, M., and Tichler, J. (1981), *"Transient climate response to changing carbon dioxide concentration"*, Climatic Change 3, 137–153.

Miller, A.S. and Mintzer, I.M. (1986), *"The sky is the limit: Strategies for protecting the ozone layer"*, World Resources Institute, Washington D.C.

Mintzer, I. (1987), *"A matter of degrees: the potential for controlling the greenhouse effect"*, World Resources Institute, Washington, D.C.

Molofsky, J., Hall, C.A.S. and Meyers, N. (1986), *"A comparison of tropical forest surveys"*, US Department of Energy, Report no. TRO32, Washington D.C.

Montalembert, M.R. de, and Clément, J. (1983), *"Fuelwood supplies in the developing countries"*, FAO, Rome.

Montgomery, D.C. (1984), *"Design and analysis of experiments"* (second edition), J. Wiley & Sons Inc., New York.

Mook, W.G., and Engelsman, F.M.R. (1983), *"The atmospheric CO_2 concentration and the role of the oceans"* (in Dutch), Chemical Magazine, December 1983, 669–671.

Myers, N. (1984), *"Conversion of tropical moist forests"*, National Academy of Sciences, Washington, D.C.

OECD: (1983), *"Long-term outlook for the world automobile industry"*, Paris.

Oerlemans, J. (1982), *"Response of the Antarctic ice sheet to a climatic warming: a model study"*, Journal of Climatology 2, 1–11.

Oerlemans, J. (1986a), *"Glaciers as indicators of a carbon dioxide warming"*, Nature 320, no. 6063, 607–609.

Oerlemans, J. (1986b), *"Greenhouse warming and changes in sea-level"*, Presented at the Technical Workshop: Developing policies for responding to future climatic change, Villach, Australia, 28 Sept. 31 Oct. 1987, 14 pp.

Oerlemans, J. (1987), *"Greenhouse warming and changes in sea-level"*, Presented at the Technical Workshop: Developing policies for responding to future climatic change, Villach, Austria, September 28 – October 31 1987, 14 pp.

Oerlemans, J. (1988), *"Simulation of historic glacier variations with a simple climate-glacier model"*, Journal of Glaciology 34, no. 118.

Oerlemans, J. (1989), *"A projection of future sea level rise"*, Climatic Change.

Oeschger, H., Siegenthaler, U., Schotterer, V., and Gugelmann, A. (1975), *"A box diffusion model to study the carbon dioxide exchange in nature"*, Tellus 27, 168–192.

Olsthoorn, T.N. (1987), *"Integrated environmental modelling, definition report"*, (in Dutch), RIVM, Bilthoven, The Netherlands.

Olsthoorn, T.N., Jaarsveld, J.A. van, Knoop, J.M., Egmond, N.D. van, Mülschlegel, J.H.C., Duijvenboden, W. van (1989), *"Integrated modelling in the Netherlands"*, paper presented at the International Conference on Environmental Models: Emissions and Consequences, Risø, Denmark, 22–25, May 1989.

Peng, T.H., Broecker, W.S., Freyer, H.D., and Trumbore, S. (1983), *"A deconvolution of the tree-ring based ^{13}C record"*, Journal of Geophysical Research 88, 3609–3620.

Pestel, E. (1988), *"Beyond the limits to growth"* (in Dutch), Report to the Club of Rome, Amsterdam, The Netherlands.

Pugh, A.L. (1983), *"DYNAMO User's manual"*, sixth edition, MIT Press, Cambridge, Massachusetts, and London.

Pulles, J.W. (1985), *"Beleidsanalyse voor de waterhuishouding in Nederland (PAWN)"*, Rijkswaterstaat, Hoofddirectie van de Waterstaat, The Hague.

Ramanathan, V., Lian, MS, Cess, RD. (1979), *"Increased atmospheric CO_2 : Zonal and Seasonal Estimates of the Effect on the Radiation Energy Balance and Surface Temperature"*, Journal of Geophysical Research 84, 4949–4958.

Ramanathan, V., Cicerone, R.J., Singh, H.B. and Kiehl, J.T. (1985), *"Trace gas trends and their potential role in climate change"*, Journal of Geophysical Research-90, 5547–5566.

Repetto, R. (1988), *"The forest for the trees? Government Policies and the Misuse of Forest Resources"*, World Resources Institute, Washington.

Response Strategies Working Group (1989a), *"Emissions scenarios of the response strategies working group of the intergovernmental panel on climate change"*, Draft Report of the U.S.-Netherlands Expert Group on Emissions Scenarios, Bilthoven, The Netherlands.

Response Strategies Working Group (1989b), *"Emissions scenarios of the response strategies working group of the intergovernmental panel on climate change"*, Draft Appendix of the U.S.-Netherlands Expert Group on Emissions Scenarios, Bilthoven, The Netherlands.

Revelle, R.R. (1983), *"Probable future changes in sea level resulting from increased atmospheric carbon dioxide"*. In: 'Changing Climate', Report of the Carbon Dioxide Assessment Committee, NRC/NAS/NAE/IOM, National Academy Press, Washington D.C.

Revelle, R.R. (1983), *"Probable future changes in sea level resulting from increased carbon dioxide"*. In: Changing Climate, National Academy Press, Washington, D.C., 433–448.

Rijkswaterstaat (1989), *"Survey technical foundation discussion paper coastal defence"* (in Dutch), Coastal defence after 1990, Technical Report 0, The Netherlands.

Robertson, G.P., and Tiedje, J.M., (1988), *"Deforestation alters denitrification in a lowland tropical rain forest"*, Nature 336, 1988, 756–759.

Robin, G.Q. de (1986), *"Changing the sea level: projecting the rise in sea level caused by warming of the atmosphere"*. In: 'The greenhouse effect, climatic change, and ecosystems', B. Bolin, B.R. Döös, J. Jäger,

R.A. Warrick (eds), SCOPE 29, Chichester, UK, John Wiley and Sons, 323–359.

Robin, G. de Q. (1986), *"Changing the sea level, in the greenhouse effect, climatic change and ecosystems"*, Bolin, B., Döös, B.R., Jäger, J. and Warrick, R.A. (eds), John Wiley and Sons, New York, 323–359.

Ronde, J.G. de (1988), *"What will happen to the Netherlands if sea level rise accelerates"*, Rijkswaterstaat.

Ronde, J.G. de, and Vrees, L. de (1990), Personal communication, Rijkswaterstaat, The Netherlands.

Ronde, J.G. de, and Vogel, J.A. (1989), *"Sea level rise, hydro meteo scenario"* (in Dutch), Coastal defence after 1990, Technical Report 6, The Netherlands.

Rotmans, J. (1986), *"The development of a simulation model for the global CO_2 problem"*, (in Dutch), Report no. 840751001, RIVM, Bilthoven, The Netherlands.

Rotmans, J. (1989), *"A scenario study of the greenhouse effect"*, In Atmospheric Ozone Research and its Policy Implications, (eds. T. Schneider, S.D. Lee, G.J.R. Wolters, L.D. Grant) 377–384, Elsevier Amsterdam.

Rotmans, J. (1990), *"Sea level rise potentials"*, in Targets and Indicators, Draft Report of Working Group II of the Advisory Group on Greenhouse Gases (AGGG), UNEP/WMO.

Rotmans, J., Boois, H. de, and Swart, R.J. (1990a), *"An integrated model for the assessment of the Greenhouse Effect: the Dutch approach"*, Climatic Change, Vol. 16, no. 3, 331–356; also printed as RIVM Report no. 758471009, Bilthoven 1989, The Netherlands.

Rotmans, J., Swart, R.J., and Vrieze, O.J. (1990b), *"The role of the CH_4-CO-OH cycle in the greenhouse problem"*, Science of the Total Environment, Vol. 94, no. 3, 233–252.

Rotmans, J., and Swart, R.J. (1990), *"Modelling tropical deforestation and its consequences for global climate"*, submitted to Ecological Modelling.

Rotmans, J., and Swart, R.J. (1990), *"Should the world phase-out fossil fuels?"*, to be published in Environmental Management.

Rotmans, J., and Elzen, M.G.J. den, (1990), *"Temperature increasing potentials (TIPS) of greenhouse gases"*, Report no. 222901001, RIVM, Bilthoven, The Netherlands.

Rotmans, J., Lashof, D.A., and Elzen, M.G.J. den, (1990c), *"Emissions, concentrations, and temperature"*, in Targets and Indicators, Draft Report of Working Group II of the Advisory Group on Greenhouse Gases (AGGG), UNEP/WMO.

Rotmans, J. and Eggink, E. (1988), *"Methane as a greenhouse gas: a simulation model of the atmospheric chemistry of the CH_4-CO-OH cycle"*, Report no. 758471002, RIVM, Bilthoven, The Netherlands.

Rotmans, J., Ham, G. van, Kleijnen, J.P.C. (1990), *"Sensitivity analysis by metamodels and experimental designs, applied on a simulation model of the greenhouse effect"*, (in Dutch), accepted by Kwantitatieve Methoden.

Rotmans, J., Vrieze O.J., Peek, G.H.J.C., Veraart, W.N.G.M. (1988), *"Experimental design and metamodelling applied to a CO_2 simulation model"*, (in Dutch), RIVM-report, Report no. 758471003, Bilthoven, the Netherlands.

Rotmans, J., and Vrieze, O.J. (1990), *"Metamodelling and experimental design: case study of the greenhouse effect"*, to be published in the European Journal of Operations Research, and is also a research report of the University of Limburg, Report M 88-03, Maastricht, the Netherlands.

Rotty, R.M. and Masters, C.D. (1985), *"Carbon dioxide from fossil fuel combustion: trends, resources and technological implications"*; In: Atmospheric Carbon Dioxide and the Global Carbon Cycle; US Department of Energy, Washington D.C.

Rotty, R.M. (1987), *"A look at 1983 CO_2 emissions from fossil fuels (with preliminary data for 1984)"*, Tellus 39, 203–208.

Sachs, J., Welch, W.J., Mitchell, T.J., and Wynn, H.P. (1989), *"Design and Analysis of computer experiments"*, Statistical Science 4, no. 4, 409–435.

Sassin, W., Jäger, J., Primio, J.C. di, and Fischer, W. (1988), *"Das Klimaproblem zwischen Naturwissenschaft und Politik"*, Kernforschungs-anlage Jährlich.

Schlesinger, M.E. (1985), *"Analysis of results from energy balance and radiative-convective models"*, in Projecting the Climatic Effects of Increasing Carbon Dioxide, US Department of Energy, DOE/ER-0237, Washington D.C.

Schlesinger, M.E. (1986), *"Equilibrium and transient climatic warming induced by increased atmospheric CO_2"*, Climate Dynamics-1, 35–51.

Schneider, S.H., and Thompson, S.L. (1981), *"Atmospheric CO_2 and climate: importance of the transient response"*, Journal of Geophysical Research 86, 3135–3147.

Schneider, S.H. (1989), *"The changing climate"*, Scientific American, September 1989, 38–47.

Schuurmans, C.J.E., Oerlemans, J., Mureau, R., Dool, H.M., van den (1982), *"Fysische aspecten van het CO_2 probleem"*, Energiespectrum, September 1982, 218–227.

Seiler, W. (1984), *"Contribution of biological processes to the global budget of CH_4 in the atmosphere"*; In: Current perspectives in microbiological ecology, Am. Soc. of Microbiology, Washington D.C.

Sheppard, J.C. et al. (1982), *"Inventory of global methane sources and their production rates"*, Journal of Geophysical Research 87, 1305–1312.

Shine, K.P., and Henderson-Sellers, A. (1983), *"Modelling climate and the nature of climate models: A review"*, Journal of Climatology 3, 81–94.

Siegenthaler, U. and Oeschger, H. (1987), *"Biospheric CO_2 emissions during the past 200 years, reconstructed by deconvolution of ice core data"*, Tellus 39B, 140–154.

Soest, J.P. van, Kasteren, J. van, Jansen, F., Groen, A., Hartman, B., and Bakker, J. (1988), *"Reality of the model"* (in Dutch), Aramith Uitgevers, Amsterdam, The Netherlands.

Spelman, M.J., and Manabe, S. (1984), *"Influence of oceanic heat transport upon the sensitivity of a model climate"*, Journal of Geophysical Research 89, 571–586.

Stewart, R.W., Hameed, S. and Pinto, J.P. (1977), *"Photochemistry of tropospheric ozone"*, Journal of Geophysical Research 82, 3134–3140.

Stive, M.J.F. (1989), *"Coastal prediction development coastline 1990–2090"*, Coastal defence after 1990, Technical Report 5, Rijkswaterstaat, The Netherlands.

Swart, R.J. (1988), *"Global anthropogenic emissions of carbon monoxide and non-methane volatile organic compounds as input for the CH_4-CO-OH cycle module; A contribution to IMAGE, the Integrated Model for the Assessment of the Greenhouse Effect"*, Report no. 758471004, RIVM, Bilthoven, The Netherlands.

Swart, R.J., Boois, H. de, and Rotmans, J. (1989), *"Targeting climate change"*, International Environmental Affairs, A Journal for Research and Policy 1, no 3, 222–234.

Swart, R.J. and Rotmans, J. (1989a), *"Food or forest? Can the tropical forests survive along with continuing growth of population and economy?"*, In: Soils and the greenhouse effect, Bouwman, A.F. (ed.), 431–439, John Wiley and Sons, Chichester.

Swart, R.J. and Rotmans, J. (1989b), *"A scenario study on causes of tropical deforestation and effects on the global carbon cycle"*, RIVM, Report no. 758471007, Bilthoven, The Netherlands.

Swart, R.J., and Rotmans, J. (1989), *"IMAGE: a tool for long-term global greenhouse policy analysis"*, in Climate and Energy: the feasibility of controlling CO_2 emissions, P.A. Okken, R.J. Swart, S. Zwerver (eds.), 18–27, Kluwer Academic Publishes, Dordrecht, The Netherlands.

Swart, R.J., and Rotmans, J. (1990a), *"Emission scenarios for the Intergovernmental Panel on Climate Change"*, (in Dutch), Milieu 1990/3.

Swart, R.J., and Rotmans, J. (1990b), *"Stabilizing atmospheric concentrations: towards international methane control"*, submitted to Ambio.

Sze, N.D. (1977), *"Anthropogenic CO emissions: implications for the atmospheric CO-OH-CH_4 cycle"*, Science 195, 673–675.

Thissen, W. (1978), *"Investigations into the club of Rome's world3 model: lessons for understanding complicated models"*, Thesis, Krips Repro, Meppel, The Netherlands.

Thissen, W. (1978), *"Investigations into the world3 model: lessons for understanding complicated models"*, IEEE Transactions on Systems, Man, and Cybernetics, vol. smc-8, no. 3.

Thomas, R.H., and Bentley, C.R. (1978), *"A model for holocene retreat of the West Antarctic Ice Sheet"*, Quartenary Res 10, 150–170.

Thompson, A.M.and Cicerone, R.J. (1986) *"Possible perturbations to atmospheric CO, CH_4 and OH "*, Journal of Geophysical Research 92, 10853–10864.

Tricot, Ch. and Berger, A. (1987), *"modelling the equilibrium and transient responses of global temperature to past and future trace gas concentrations"*, Climate Dynamics 2, 39–61.

United Nations (1986), *"World population prospects, estimates and projections as assessed in 1984"*, Population Studies 98.

United Nations Environment Program (1987), *"Montreal protocol on substances that deplete the ozone layer: final act"*, Montreal.

United Nations Environment Program, and World Meteorological Organization (1989), *"Scientific assessment of stratospheric ozone"*, Chapter 4: ODPs and GWPs.

United States Department of Energy (1985a), *"Atmospheric carbon dioxide and the global carbon cycle"*, J.R. Trabalka (ed.), USDOE, DOE/ER/-0239-1, Washington D.C.

United States Department of Energy (1985b), *"Methods of uncertainty analysis for a global carbon dioxide model"*, R.H. Gardner and J.R. Trabalka (eds.), USDOE, DOE/OR/21400-4, Washington D.C.

United States Department of Energy (1985c), *"Glaciers, ice sheets, and sea level: effect of a CO_2 induced climatic change"*, USDOE, DOE/ER/60235-1, Washington D.C.

Veen, C.J. van der, (1986), *"Ice sheets, atmospheric CO_2 and sea level"*, Thesis, University of Utrecht, the Netherlands.

Veen, C.J. van der (1986), *"Ice sheets, atmospheric CO_2 and sea level"*, Thesis, University of Utrecht, The Netherlands, 184 pp.

Volz, A., Ehhalt, D.H. and Derwent, R.G. (1981), *"Seasonal and latitudinal variation of CO and the tropospheric concentration of OH radicals"*, Journal of Geophysical Research 86, 5163–5171.

Vrijling, J.K. (1985), *"Enkele gedachten over een aanvaardbaar risiconiveau in Nederland"*, Technische Adviescommissie van de Waterkering, Government printing office, The Hague.

Wang, W.-C., and Molnar, G. (1985), *"A model study of the Greenhouse Effect due to increasing atmospheric CH_4, N_2O, CF_2Cl_2, and $CFCl_3$"*, Journal of Geophysical Research 90, 12971-12980-12980.

Washington, W.M., and Parkinson, C.L. (1986), *"An introduction to three-dimensional climate modelling"*, Oxford University Press, Oxford, England.

Watts, J.A. (1982), *"The carbon dioxide question: data sampler"*, in the Carbon Dioxide Review, W.C. Clark (ed.), 457–460.

Weiss, R.F. (1981), *"The temporal and spatial distribution of tropospheric nitrous oxide"*, Journal of Geophysical Research 86, 7185–7195.

Wiersum, K.F., and Ketner, P. (1989), *"Reforestation, a feasible contribution to reducing the atmospheric carbon dioxide content"*, in Climate and Energy, the Feasibility of Controlling CO_2 Emissions', P.A. Okken, R.J. Swart, and S. Zwerver (eds.), Kluwer Academic Publishers, Dordrecht, The Netherlands.

Wigley, T.M.L. (1985), *"Carbon dioxide, trace gases and global warming"*, Climate Monitor 13, no. 5, 132–148.

Wigley, T.M.L. (1987), *"Relative contributions of different trace gases to the greenhouse effect"*, Climate monitor-16, no. 1, 14–28.

Wigley, T.M.L. (1988), *"Future CFC concentrations under the Montreal Protocol and their greenhouse effect implications"*, Nature 335, 333–335.

Wigley, T.M.L., Raper, S.C.B. (1987), "*Thermal expansion of sea water associated with global warming*", Nature 330, 127–131.

Wigley, T.M.L. and Schlesinger, M.E. (1985), "*Analytical solution for the effect of increasing CO_2 on global mean temperature*", Nature 315, 649–652.

Wind, H.G. (1987), "*Impact of sea level rise on society*", Balkema, Rotterdam, The Netherlands.

Wolman, M.G. and Fournier, F.G.A. (ed.) (1987), "*Land transformation in agriculture*", SCOPE-report 32, Wiley.

World Resources Institute/Institute for Environment and Development (1986), "*World Resources 1986*", Washington D.C.

World Resources Institute/Institute for Environment and Development (1987), "*World Resources 1987*", Washington D.C.

World Resources Institute/Institute for Environment and Development (1988/1989): "*World Resources 1988/1989*", Washington D.C.

Wuebbles, D. (1981), "*The relative efficiency of a number of halocarbons for destroying stratospheric ozone*", Report, UCID-18924, Lawrence Livermore National Laboratory, Livermore.

Wuebbles, D.J. (1983), "*Chlorocarbon emission scenarios: potential impact on stratospheric ozone*", Journal of Geophysical Research 88, 1433–1443.

Yearsley, J.R., and Lettenmaier, D.P. (1987), "*Model complexity and data worth: an assessment of changes in the global carbon budget*", Ecological Modelling 39, 201–226.